# Variable-length Codes for Data Compression

David Salomon

# Variable-length Codes
# for Data Compression

Springer

Professor David Salomon (emeritus)
Computer Science Department
California State University
Northridge, CA 91330-8281
USA
email: david.salomon@csun.edu

British Library Cataloguing in Publication Data
A catalogue record for this book is available from the British Library

Library of Congress Control Number:

ISBN 978-1-84628-958-3          ISBN 978-1-84628-959-0 (eBook)
DOI 10.1007/978-1-84628-959-0
Printed on acid-free paper.

9 8 7 6 5 4 3 2 1

Springer Science+Business Media
springer.com

*To the originators and developers of the codes,*
*Apostolico, Capocelli, Elias, Fenwick, Fraenkel,*
*Golomb, Huffman, Klein, Pigeon, Rice, Stout,*
*Tsai, Tunstall, Villasenor, Wang, Wen, Wu,*
*Yamamoto, and others.*

To produce a mighty book, you must choose a mighty theme.
—Herman Melville

# Preface

The dates of most of the important historical events are known, but not always very precisely. We know that Kublai Khan, grandson of Ghengis Khan, founded the Yuan dynasty in 1280 (it lasted until 1368), but we don't know precisely (i.e., the month, day and hour) when this act took place. A notable exception to this state of affairs is the modern age of telecommunications, a historical era whose birth is known precisely, up to the minute. On Friday, 24 May 1844, at precisely 9:45 in the morning, Samuel Morse inaugurated the age of modern telecommunications by sending the first telegraphic message in his new code. The message was sent over an experimental line funded by the American Congress from the Supreme Court chamber in Washington, DC to the B & O railroad depot in Baltimore, Maryland. Taken from the Bible (*Numbers* 23:23), the message was "What hath God wrought?" It had been suggested to Morse by Annie Ellsworth, the young daughter of a friend. It was prerecorded on a paper tape, was sent to a colleague in Baltimore, and was then decoded and sent back by him to Washington. An image of the paper tape can be viewed at [morse-tape 06].

Morse was born near Boston and was educated at Yale. We would expect the inventor of the telegraph (and of such a sophisticated code) to have been a child prodigy who tinkered with electricity and gadgets from an early age (the electric battery was invented when Morse was nine years old). Instead, Morse became a successful portrait painter with more than 300 paintings to his credit. It wasn't until 1837 that the 46-year-old Morse suddenly quit his painting career and started thinking about communications and tinkering with electric equipment. It is not clear why he made such a drastic career change at such an age, but it is known that two large, wall-size paintings that he made for the Capitol building in Washington, DC were ignored by museum visitors and rejected by congressmen. It may have been this disappointment that gave us the telegraph and the Morse code.

Given this background, it is easy to imagine how the 53-year-old Samuel Morse felt on that fateful day, Friday, 24 May 1844, as he sat hunched over his mysterious apparatus, surrounded by a curious crowd of onlookers, some of whom had only a vague idea of what he was trying to demonstrate. He must have been very anxious, because his telegraph project, his career, and his entire future depended on the success of this one test. The year before, the American Congress awarded him $30,000 to prepare this historical test and prove the value of the electric telegraph (and thus also confirm the ingenuity of yankees), and here he is now, dependent on the vagaries of his batteries, on the new, untested 41-mile-long telegraph line, and on a colleague in Baltimore.

Fortunately, all went well. The friend in Baltimore received the message, decoded it, and resent it within a few minutes, to the great relief of Morse and to the amazement of the many congressmen assembled around him.

The Morse code, with its quick dots and dashes (Table 1), was extensively used for many years, first for telegraphy, and beginning in the 1890s, for early radio communications. The development of more advanced communications technologies in the 20th century displaced the Morse code, which is now largely obsolete. Today, it is used for emergencies, for navigational radio beacons, land mobile transmitter identification, and by continuous wave amateur radio operators.

| | | | | | | | |
|---|---|---|---|---|---|---|---|
| A | .- | N | -. | 1 | .---- | Period | .-.-.- |
| B | -... | O | --- | 2 | ..--- | Comma | --..-- |
| C | -.-. | P | .--. | 3 | ...-- | Colon | ---... |
| Ch | ---- | Q | --.- | 4 | ....- | Question mark | ..--.. |
| D | -.. | R | .-. | 5 | ..... | Apostrophe | .----. |
| E | . | S | ... | 6 | -.... | Hyphen | -....- |
| F | ..-. | T | - | 7 | --... | Dash | -..-. |
| G | --. | U | ..- | 8 | ---.. | Parentheses | -.--.- |
| H | .... | V | ...- | 9 | ----. | Quotation marks | .-..-. |
| I | .. | W | .-- | 0 | ----- | | |
| J | .--- | X | -..- | | | | |
| K | -.- | Y | -.-- | | | | |
| L | .-.. | Z | --.. | | | | |
| M | -- | | | | | | |

Table 1: The Morse Code for English.

Our interest in the Morse code is primarily with a little-known aspect of this code. In addition to its advantages for telecommunications, the Morse code is also an early example of text compression. The various dot-dash codes developed by Morse (and possibly also by his associate, Alfred Vail) have different lengths, and Morse intuitively assigned the short codes (a single dot and a single dash) to the letters E and T, the longer, four dots-dashes, he assigned to Q, X, Y, and Z. The even longer, five dots-dashes codes, were assigned to the 10 digits, and the longest codes (six dots and dashes) became those of the punctuation marks. Morse also specified that the signal for error is eight consecutive dots, in response to which the receiving operator should delete the last word received.

It is interesting to note that Morse was not the first to think of compression (in terms of time saving) by means of a code. The well-known Braille code for the blind was developed by Louis Braille in the 1820s and is still in common use today. It consists of groups (or cells) of 3 × 2 dots each, embossed on thick paper. Each of the six dots in a group may be flat or raised, implying that the information content of a group is equivalent to six bits, resulting in 64 possible groups. The letters, digits, and common punctuation marks do not require all 64 codes, which is why the remaining groups may be used to code common words—such as **and**, **for**, and **of**—and common strings of letters—such as **ound**, **ation**, and **th**.

The Morse code has another feature that makes it relevant to us. Because the individual codes have different lengths, there must be a way to identify the end of a code. Morse solved this problem by requiring accurate relative timing. If the duration of a dot is taken to be one unit, then that of a dash is three units, the space between the dots and dashes of one character is one unit, the space between characters is three units, and the interword space is six units (five for automatic transmission). This book is concerned with the use of variable-length codes to compress digital data. With these codes, it is important not to have any extra spaces. In fact, there is no such thing as a space, because computers use only zeros and 1's. Thus, when a string of data symbols is compressed by assigning short codes (that are termed "codewords") to the symbols, the codewords (whose lengths vary) are concatenated into a long binary string without any spaces or separators. Such variable-length codes must therefore be designed to allow for unambiguous reading. Somehow, the decoder should be able to read bits and identify the end of each codeword. Such codes are referred to as uniquely decodable or uniquely decipherable (UD).

Variable-length codes have become important in many areas of computer science. This book is a survey of this important topic. It presents the principles underlying this type of codes and describes the important classes of variable-length codes. Many examples illustrate the applications of these codes to data compression. The book is devoted to the codes, which is why it describes very few actual compression algorithms. Notice that many important (and some not so important) methods, algorithms, and techniques for compressing data are described in detail in [Salomon 06].

The term *representation* is central to our discussion. A number can be represented in decimal, binary, or any other number base (or number system, see Section 2.18). Mathematically, a representation is a bijection (or a bijective function) of an infinite, countable set $S_1$ of strings onto another set $S_2$ of strings (in practice, $S_2$ consists of binary strings, but it may also be ternary or based on other number systems), such that any concatenation of any elements of $S_2$ is UD. The elements of $S_1$ are called data symbols and those of $S_2$ are codewords. Set $S_1$ is an alphabet and set $S_2$ is a code. An interesting example is the standard binary notation. We normally refer to it as the binary representation of the integers, but according to the definition above it is not a representation because it is not UD. It is easy to see, for example, that a string of binary codewords that starts with 11 can be either two consecutive 1's or the code of 3.

---

A function $f : X \Rightarrow Y$ is said to be bijective, if for every $y \in Y$, there is exactly one $x \in X$ such that $f(x) = y$.

Figure 3.19 and Table 3.22 list several variable-length UD codes assigned to the 26 letters of the English alphabet.

This book is aimed at readers who have a basic knowledge of data compression and who want to know more about the specific codes used by the various compression algorithms. The necessary mathematical background includes logarithms, polynomials, a bit of calculus and linear algebra, and the concept of probability. This book is not intended as a guide to software implementors and has no programs. Errors, mistypes, comments, and questions should be sent to the author's email address below.

It is my pleasant duty to acknowledge the substantial help and encouragement I have received from Giovanni Motta and Cosmin Truţa and for their painstaking efforts. They read drafts of the text, found many errors and misprints, and provided valuable comments and suggestions that improved this book and made it what it is. Giovanni also wrote part of Section 2.12.

---

If, by any chance, I have omitted anything more or less proper or necessary, I beg forgiveness, since there is no one who is without fault and circumspect in all matters.
　　　　　　　　　　　　　　　　　　—Leonardo Fibonacci, *Libe Abaci* (1202)

---

`dsalomon@csun.edu`                                           David Salomon

The Preface is the most important part of
the book. Even reviewers read a preface.
　　　　　　　　　　　　—Philip Guedalla

# Contents

# Contents

An adequate table of contents serves as a synopsis or headline display
of the design or structural pattern of the body of the report.

—D. E. Scates and C. V. Good, *Methods of Research*

# Introduction

The discipline of data compression has its origins in the 1950s and 1960s and has experienced rapid growth in the 1980s and 1990s. Currently, data compression is a vast field encompassing many approaches and techniques. A student of this field realizes quickly that the various compression algorithms in use today are based on and require knowledge of diverse physical and mathematical concepts and topics, some of which are included in the following, incomplete list: Fourier transforms, finite automata, Markov processes, the human visual and auditory systems—statistical terms, distributions, and concepts—Unicode, XML, convolution, space-filling curves, Voronoi diagrams, interpolating polynomials, Fibonacci numbers, polygonal surfaces, data structures, the Vandermonde determinant, error-correcting codes, fractals, the Pascal triangle, fingerprint identification, and analog and digital video.

Faced with this complexity, I decided to try and classify in this short introduction most (but not all) of the approaches to data compression in four classes as follows: (1) block-to-block codes, (2) block-to-variable codes, (3) variable-to-block codes, and (4) variable-to-variable codes (the term "fixed" is sometimes used instead of "block"). Other approaches to compression, such as mathematical transforms (orthogonal or wavelet) and the technique of arithmetic coding, are not covered here. Following is a short description of each class.

■ Block-to-block codes constitute a class of techniques that input $n$ bits of raw data at a time, perform a computation, and output the same number of bits. Such a process results in no compression; it only transforms the data from its original format to a format where it becomes easy to compress. Thus, this class consists of transforms. The discrete wavelet, discrete cosine, and linear prediction are examples of transforms that are commonly used as the first step in the compression of various types of data. Here is a short description of linear prediction.

Audio data is common in today's computers. We all have mp3, FLAC, and other types of compressed audio files in our computers. A typical lossless audio compression technique consists of three steps. (1) The original sound is sampled (digitized). (2) The audio samples are converted, in a process that employs linear prediction, to small numbers called residues. (3) The residues are replaced by variable-length codes. The last step is the only one that produces compression.

Linear prediction of audio samples is based on the fact that most audio samples are similar to their near neighbors. One second of audio is normally converted to many thousands of audio samples (44,100 samples per second is typical), and adjacent samples tend to be similar because sound rarely varies much in pitch or frequency during one second. If we denote the current audio sample by $s(t)$, then linear prediction computes a predicted value $\hat{s}(t)$ from the $p$ immediately-preceding samples by a linear combination of the form

$$\hat{s}(t) = \sum_{i=1}^{p} a_i s(t-i).$$

Parameter $p$ depends on the specific algorithm and many also be user controlled. Parameters $a_i$ are linear coefficients that are also determined by the algorothm.

If the prediction is done properly, the difference (which is termed residue or residual) $e(t) = s(t) - \hat{s}(t)$ will almost always be a small (positive or negative) number, although in principle it could be about as large as $s(t)$ or $-s(t)$. The difference between the various linear prediction methods is in the number $p$ of previous samples that they employ and in the way they determine the linear coefficients $a_i$.

■   Block-to-variable codes are the most important of the four types discussed here. Each symbol of the input alphabet is assigned a variable-length code according to its frequency of occurrence (or, equivalently, its probability) in the data. Compression is achieved if short codes are assigned to commonly-occurring (high probability) symbols and long codes are assigned to rare symbols. Many statistical compression methods employ this type of coding, most notably the Huffman method (Section 1.13). The difference between the various methods is mostly in how they compute or estimate the probabilities of individual data symbols. There are three approaches to this problem, namely static codes, a two-pass algorithm, and adaptive methods.

Static codes. It is possible to construct a set of variable-length codes and permanently assign each code to a data symbol. The result is a static code table that is built into both encoder and decoder. To construct such a table, the developer has to analyze large quantities of data and determine the probability of each symbol. For example, someone who develops a compression method that employs this approach to compress text, has to start by selecting a number of representative "training" documents, count the number of times each text character appears in those documents, compute frequencies of occurrence, and use this fixed, static statistical model to assign variable-length codewords to the individual characters. A compression method based on a static code table is simple, but the results (the compression ratio for a given text file) depend on how much the data resembles the statistics of the training documents.

A two-pass algorithm. The idea is to read the input data twice. The first pass simply counts symbol frequencies and the second pass performs the actual compression by replacing each data symbol with a variable-length codeword. In between the two passes, the code table is constructed by utilizing the symbols' frequencies in the particular data being compressed (the statistical model is taken from the data itself). Such a method features very good compression, but is slow because reading a file from an input device, even a fast disk, is slower than memory-based operations. Also, the code table is constructed individually for each data file being compressed, so it has to be included in the compressed file, for the decoder's use. This reduces the compression ratio but not significantly, because a code table typically contains one variable-length code for each of the 128 ASCII characters or for each of the 256 8-bit bytes, so its total length is only a few hundred bytes.

An adaptive method starts with an empty code table, or with a tentative table, and modifies the table as more data is read and processed. Initially, the codes assigned to the data symbols are inappropriate and are not based on the (unknown) probabilities of the data symbols. But as more data is read, the encoder acquires better statistics of the data and exploits it to improve the codes (the statistical model adapts itself gradually to the data that is being read and compressed). Such a method has to be designed to permit the decoder to mimic the operations of the encoder and modify the code table in lockstep with it.

A simple statistical model assigns variable-length codes to symbols based on symbols' probabilities. It is possible to improve the compression ratio significantly by basing the statistical model on probabilities of pairs or triplets of symbols (digrams and trigrams), instead of probabilities of individual symbols. The result is an $n$-order statistical compression method where the previous $n$ symbols are used to predict (i.e., to assign a probability to) the current symbol. The PPM (prediction by partial matching) and DMC (dynamic Markov coding) methods are examples of this type of algorithm.

It should be noted that arithmetic coding, an important statistical compression method, is included in this class, but operates differently. Instead of assigning codes to individual symbols (bits, ASCII codes, Unicodes, bytes, etc.), it assigns one long code to the entire input file.

■ Variable-to-block codes is a term that refers to a large group of compression techniques where the input data is divided into chunks of various lengths and each chunk of data symbols is encoded by a fixed-size code. The most important members of this group are run-length encoding and the various LZ (dictionary-based) compression algorithms.

A dictionary-based algorithm saves bits and pieces of the input data in a special buffer called a dictionary. When the next item is read from the input file, the algorithm tries to locate it in the dictionary. If the item is found in the dictionary, the algorithm outputs a token with a pointer to the item plus other information such as the length of the item. If the item is not in the dictionary, the algorithm adds it to the dictionary (based on the assumption that once an item has appeared in the input, it is likely that it will appear again) and outputs the item either in raw format or as a special, literal token. Compression is achieved if a large item is replaced by a short token. Quite a few dictionary-based algorithms are currently known. They have been developed by many scientists and researchers, but are all based on the basic ideas and pioneering work of Jacob Ziv and Abraham Lempel, described in [Ziv and Lempel 77] and [Ziv and Lempel 78].

A well-designed dictionary-based algorithm can achieve high compression because a given item tends to appear many times in a data file. In a text file, for example, the same words and phrases may appear many times. Words that are common in the language and phrases that have to do with the topic of the text, tend to appear again and again. If they are kept in the dictionary, then more and more phrases can be replaced by tokens, thereby resulting in good compression.

The differences between the various LZ dictionary methods are in how the dictionary is organized and searched, in the format of the tokens, in the way the algorithm handles items not found in the dictionary, and in the various improvements it makes to the basic method. The many variants of the basic LZ approach employ improving techniques such as a circular buffer, a binary search tree, variable-length codes or dynamic Huffman coding to encode the individual fields of the token, and other tricks of the programming trade. Sophisticated dictionary organization eliminates duplicates (each data symbol is stored only once in the dictionary, even if it is part of several items), implements fast search (binary search or a hash table instead of slow linear search), and may discard unused items from time to time in order to regain space.

The other important group of variable-to-block codes is run-length encoding (RLE). We know that data can be compressed because the common data representations are redundant, and one type of redundancy is runs of identical symbols. Text normally does not feature long runs of identical characters (the only examples that immediately come to mind are runs of spaces and of periods), but images, especially monochromatic (black and white) images, may have long runs of identical pixels. Also, an audio file may have silences, and even one-tenth of second worth of silence typically translates to 4,410 identical audio samples.

A typical run-length encoder identifies runs of the same symbol and replaces each run with a token that includes the symbol and the length of the run. If the run is shorter than a token, the raw symbols are output, but the encoder has to make sure that the decoder can distinguish between tokens and raw symbols.

Since runs of identical symbols are not common in many types of data, run-length encoding is often only one part of a larger, more sophisticated compression algorithm.

■    Variable-to-variable codes is the general name used for compression methods that select variable-length chunks of input symbols and compress each chunk by replacing it with a variable-length code.

A simple example of variable-to-variable codes is run-length encoding combined with Golomb codes, especially when the data to be compressed is binary. Imagine a long string of 0's and 1's where one value (say, 0) occurs more often than the other value. This value is referred to as the more probable symbol (MPS), while the other value becomes the LPS. Such a string tends to have runs of the MPS and Section 2.23 shows that the Golomb codes are the best candidate to compress such runs. Each run has a different length, and the various Golomb codewords also have different lengths, turning this application into an excellent example of variable-to-variable codes.

Other examples of variable-to-variable codes are hybrid methods that consist of several parts. A hybrid compression program may start by reading a chunk of input and looking it up in a dictionary. If a match is found, the chunk may be replaced by a token, which is then further compressed (in another part of the program) by RLE or variable-length codes (perhaps Huffman or Golomb). The performance of such a program may not be spectacular, but it may produce good results for many different types of data. Thus, hybrids tend to be general-purpose algorithms that can deal successfully with text, images, video, and audio data.

This book starts with several introductory sections (Sections 1.1 through 1.6) that discuss information theory concepts such as entropy and redundancy, and concepts that are used throughout the text, such as prefix codes, complete codes, and universal codes.

The remainder of the text deals mostly with block-to-variable codes, although its first part deals with the Tunstall codes and other variable-to-block codes. It concentrates on the codes themselves, not on the compression algorithms. Thus, the individual sections describe various variable-length codes and classify them according to their structure and organization. The main techniques employed to design variable-length codes are the following:

■ The phased-in codes (Section 1.9) are a slight extension of fixed-size codes and may contribute a little to the compression of a set of consecutive integers by changing the representation of the integers from fixed $n$ bits to either $n$ or $n - 1$ bits (recursive phased-in codes are also described).

■ Self-delimiting codes. These are intuitive variable-length codes—mostly due to Gregory Chaitin, the originator of algorithmic information theory—where a code signals its end by means of extra flag bits. The self-delimiting codes of Section 1.12 are inefficient and are not used in practice.

■ Prefix codes. Such codes can be read unambiguously (they are uniquely decodable, or UD codes) from a long string of codewords because they have a special property (the prefix property) which is stated as follows: Once a bit pattern is assigned as the code of a symbol, no other codes can start with that pattern. The most common example of prefix codes are the Huffman codes (Section 1.13). Other important examples are the unary, start-step-stop, and start/stop codes (Sections 2.2 and 2.3, respectively).

■ Codes that include their own length. One way to construct a UD code for the integers is to start with the standard binary representation of an integer and prepend to it its length $L_1$. The length may also have variable length, so it has to be encoded in some way or have *its* length $L_2$ prepended. The length of an integer $n$ equals approximately $\log n$ (where the logarithm base is the same as the number base of $n$), which is why such methods are often called logarithmic ramp representations of the integers. The most common examples of this type of codes are the Elias codes (Section 2.4), but other types are also presented. They include the Levenstein code (Section 2.5), Eve–Rodeh code (Section 2.6), punctured Elias codes (Section 2.7), the ternary comma code (Section 2.9), Stout codes (Section 2.11), Boldi–Vigna (zeta) codes (Section 2.12), and Yamamoto's recursive code (Section 2.13).

■ Suffix codes (codes that end with a special flag). Such codes limit the propagation of an error and are therefore robust. An error in a codeword affects at most that codeword and the one or two codewords following it. Most other variable-length codes sacrifice data integrity to achieve short codes, and are fragile because a single error can propagate indefinitely through a sequence of concatenated codewords. The taboo codes of Section 2.15 are UD because they reserve a special string (the taboo) to indicate the end of the code. Wang's flag code (Section 2.16) is also included in this category.

Note. The term "suffix code" is ambiguous. It may refer to codes that end with a special bit pattern, but it also refers to codes where no codeword is the suffix of another codeword (the opposite of prefix codes). The latter meaning is used in Section 3.5, in connection with bidirectional codes.

■    Flag codes. A true flag code differs from the suffix codes in one interesting aspect. Such a code may include the flag inside the code, as well as at its right end. The only example of a flag code is Yamamoto's code, Section 2.17.

■    Codes based on special number bases or special number sequences. We normally use decimal numbers, and computers use binary numbers, but any integer greater than 1 can serve as the basis of a number system and so can noninteger (real) numbers. It is also possible to construct a sequence of numbers (real or integer) that act as weights of a numbering system. The most important examples of this type of variable-length codes are the Fibonacci (Section 2.19), Goldbach (Section 2.21), and additive codes (Section 2.22).

■    The Golomb codes of Section 2.23 are designed in a special way. An integer parameter $m$ is selected and is used to encode an arbitrary integer $n$ in two steps. In the first step, two integers $q$ and $r$ (for quotient and remainder) are computed from $n$ such that $n$ can be fully reconstructed from them. In the second step, $q$ is encoded in unary and is followed by the binary representation of $r$, whose length is implied by the parameter $m$. The Rice code of Section 2.24 is a special case of the Golomb codes where $m$ is an integer power of 2. The subexponential code (Section 2.25) is related to the Rice codes.

■    Codes ending with "1" are the topic of Section 2.26. In such a code, all the codewords end with a 1, a feature that makes them the natural choice in special applications.

■    Variable-length codes are designed for data compression, which is why implementors select the shortest possible codes. Sometimes, however, data reliability is a concern, and longer codes may help detect and isolate errors. Thus, Chapter 3 discusses robust codes. Section 3.3 presents synchronous prefix codes. These codes are useful in applications where it is important to limit the propagation of errors. Bidirectional (or reversible) codes (Sections 3.5 and 3.6) are also designed for increased reliability by allowing the decoder to read and identify codewords either from left to right or in reverse.

The following is a short discussion of terms that are commonly used in this book.

■    Source. A source of data items can be a file stored on a disk, a file that is input from outside the computer, text input from a keyboard, or a program that generates data symbols to be compressed or processed in some way. In a memoryless source, the probability of occurrence of a data symbol does not depend on its context. The term i.i.d. (independent and identically distributed) refers to a set of sources that have the same probability distribution and are mutually independent.

■    Alphabet. This is the set of symbols that an application has to deal with. An alphabet may consist of the 128 ASCII codes, the 256 8-bit bytes, the two bits, or any other set of symbols.

■    Random variable. This is a function that maps the results of random experiments to numbers. For example, selecting many people and measuring their heights is a random variable. The number of occurrences of each height can be used to compute the probability of that height, so we can talk about the probability distribution of the random variable (the set of probabilities of the heights). A special important case is a discrete random variable. The set of all values that such a variable can assume is finite or countably infinite.

- Compressed stream (or encoded stream). A compressor (or encoder) compresses data and generates a compressed stream. This is often a file that is written on a disk or is stored in memory. Sometimes, however, the compressed stream is a string of bits that are transmitted over a communications line.

- The acronyms MSB and LSB refer to most-significant-bit and least-significant-bit, respectively.

- The notation $1^i0^j$ indicates a bit string of $i$ consecutive 1's followed by $j$ zeros.

> Understanding is, after all, what science is all about—and science
> is a great deal more than mere mindless computation.
>
> —Roger Penrose, *Shadows of the Mind* (1996)

# 1

# Basic Codes

The discussion in this chapter starts with codes, prefix codes, and information theory concepts. This is followed by a description of basic codes such as variable-to-block codes, phased-in codes, and the celebrated Huffman code.

## 1.1 Codes, Fixed- and Variable-Length

A code is a symbol that stands for another symbol. At first, this idea seems pointless. Given a symbol $S$, what is the use of replacing it with another symbol $Y$? However, it is easy to find many important examples of the use of codes. Here are a few.

- Any language and any system of writing are codes. They provide us with symbols $S$ that we use in order to express our thoughts $Y$.

- Acronyms and abbreviations can be considered codes. Thus, the string IBM is a symbol that stands for the much longer symbol "International Business Machines" and the well-known French university École Supérieure D'électricité is known to many simply as Supélec.

- Cryptography is the art and science of obfuscating messages. Before the age of computers, a message was typically a string of letters and was encrypted by replacing each letter with another letter or with a number. In the computer age, a message is a binary string (a bitstring) in a computer, and it is encrypted by replacing it with another bitstring, normally of the same length.

- Error control. Messages, even secret ones, are often transmitted along communications channels and may become damaged, corrupted, or garbled on their way from transmitter to receiver. We often experience low-quality, garbled telephone conversations. Even experienced pharmacists often find it difficult to read and understand a handwritten prescription. Computer data stored on magnetic disks may become corrupted because of exposure to magnetic fields or extreme temperatures. Music and movies recorded on optical discs (CDs and DVDs) may become unreadable because of scratches. In all these cases, it helps to augment the original data with error-control codes. Such codes—formally titled channel codes, but informally known as error-detecting or error-correcting codes—employ redundancy to detect and even correct, certain types of errors.

∎   ASCII and Unicode. These are character codes that make it possible to store characters of text as bitstrings in a computer. The ASCII code, which dates back to the 1960s, assigns 7-bit codes to 128 characters including 26 letters (upper- and lowercase), the 10 digits, certain punctuation marks, and several control characters. The Unicode project assigns 16-bit codes to many characters, and has a provision for even longer codes. The long codes make it possible to store and manipulate many thousands of characters, taken from many languages and alphabets (such as Greek, Cyrillic, Hebrew, Arabic, and Indic), and including punctuation marks, diacritics, mathematical symbols, technical symbols, arrows, and dingbats.

The last example illustrates the use of codes in the field of computers and computations. Mathematically, a code is a mapping. Given an alphabet of symbols, a code maps individual symbols or strings of symbols to codewords, where a codeword is a string of bits, a bitstring. The process of mapping a symbol to a codeword is termed encoding and the reverse process is known as decoding.

Codes can have a fixed or variable length, and can be static or adaptive (dynamic). A static code is constructed once and never changes. ASCII and Unicode are examples of such codes. A static code can also have variable length, where short codewords are assigned to the commonly-occurring symbols. A variable-length, static code is normally designed based on the probabilities of the individual symbols. Each type of data has different probabilities and may benefit from a different code. The Huffman method (Section 1.13) is an example of an excellent variable-length, static code that can be constructed once the probabilities of all the symbols in the alphabet are known. In general, static codes that are also variable-length can match well the lengths of individual codewords to the probabilities of the symbols. Notice that the code table must normally be included in the compressed file, because the decoder does not know the symbols' probabilities (the model of the data) and so has no way to construct the codewords independently.

A dynamic code varies over time, as more and more data is read and processed and more is known about the probabilities of the individual symbols. The dynamic (adaptive) Huffman algorithm [Salomon 06] is an example of such a code.

Fixed-length codes are known as block codes. They are easy to implement in software. It is easy to replace an original symbol with a fixed-length code, and it is equally easy to start with a string of such codes and break it up into individual codes that are then replaced by the original symbols.

There are cases where variable-length codes (VLCs) have obvious advantages. As their name implies, VLCs are codes that have different lengths. They are also known as variable-size codes. A set of such codes consists of short and long codewords. The following is a short list of important applications where such codes are commonly used.

∎   Data compression (or source coding). Given an alphabet of symbols where certain symbols occur often in messages, while other symbols are rare, it is possible to compress messages by assigning short codes to the common symbols and long codes to the rare symbols. This is an important application of variable-length codes.

■ The Morse code for telegraphy, originated in the 1830s by Samuel Morse and Alfred Vail, employs the same idea. It assigns short codes to commonly-occurring letters (the code of E is a dot and the code of T is a dash) and long codes to rare letters and punctuation marks (--.- to Q, --.. to Z, and --..-- to the comma).

■ Processor design. Part of the architecture of any computer is an instruction set and a processor that fetches instructions from memory and executes them. It is easy to handle fixed-length instructions, but modern computers normally have instructions of different sizes. It is possible to reduce the overall size of programs by designing the instruction set such that commonly-used instructions are short. This also reduces the processor's power consumption and physical size and is especially important in embedded processors, such as processors designed for digital signal processing (DSP).

■ Country calling codes. ITU-T recommendation E.164 is an international standard that assigns variable-length calling codes to many countries such that countries with many telephones are assigned short codes and countries with fewer telephones are assigned long codes. These codes also obey the prefix property (Section 1.2) which means that once a calling code $C$ has been assigned, no other calling code will start with $C$.

■ The International Standard Book Number (ISBN) is a unique number assigned to a book, to simplify inventory tracking by publishers and bookstores. The ISBN numbers are assigned according to an international standard known as ISO 2108 (1970). One component of an ISBN is a country code, that can be between one and five digits long. This code also obeys the prefix property. Once $C$ has been assigned as a country code, no other country code will start with $C$.

■ VCR Plus+ (also known as G-Code, VideoPlus+, and ShowView) is a prefix, variable-length code for programming video recorders. A unique number, a VCR Plus+, is computed for each television program by a proprietary algorithm from the date, time, and channel of the program. The number is published in television listings in newspapers and on the Internet. To record a program on a VCR, the number is located in a newspaper and is typed into the video recorder. This programs the recorder to record the correct channel at the right time. This system was developed by Gemstar-TV Guide International [Gemstar 06].

---

I gave up on new poetry myself thirty years ago, when most of it began to read like coded messages passing between lonely aliens on a hostile world.
　　　　　　　　　　　　　　　　　　　　　　　　　　　　—Russell Baker

# 1.2 Prefix Codes

Encoding a string of symbols $a_i$ with VLCs is easy. No special methods or algorithms are needed. The software reads the original symbols $a_i$ one by one and replaces each $a_i$ with its binary, variable-length code $c_i$. The codes are concatenated to form one (normally long) bitstring. The encoder either includes a table with all the pairs $(a_i, c_i)$ or it executes a procedure to compute code $c_i$ from the bits of symbol $a_i$.

Decoding is slightly more complex, because of the different lengths of the codes. When the decoder reads the individual bits of VLCs from a bitstring, it has to know either how long each code is or where each code ends. This is why a set of variable-length codes has to be carefully chosen and why the decoder has to be taught about the codes. The decoder either has to have a table of all the valid codes, or it has to be told how to identify valid codes.

We start with a simple example. Given the set of four codes $a_1 = 0$, $a_2 = 01$, $a_3 = 011$, and $a_4 = 111$ we easily encode the message $a_2 a_3 a_3 a_1 a_2 a_4$ as the bitstring $01|011|011|0|01|111$. This string can be decoded unambiguously, but not easily. When the decoder inputs a 0, it knows that the next symbol is either $a_1$, $a_2$, or $a_3$, but the decoder has to input more bits to find out how many 1's follow the 0 before it can identify the next symbol. Similarly, given the bitstring $011 \ldots 111$, the decoder has to read the entire string and count the number of consecutive 1's before it finds out how many 1's (zero, one, or two 1's) follow the single 0 at the beginning. We say that such codes are not instantaneous.

In contrast, the following set of VLCs $a_1 = 0$, $a_2 = 10$, $a_3 = 110$, and $a_4 = 111$ is similar and is also instantaneous. Given a bitstring that consists of these codes, the decoder reads consecutive 1's until it has read three 1's (an $a_4$) or until it has read another 0. Depending on how many 1's precede the 0 (zero, one, or two 1's), the decoder knows whether the next symbol is $a_1$, $a_2$, or $a_3$. The 0 acts as a separator, which is why instantaneous codes are also known as comma codes. The rules that drive the decoder can be considered a finite automaton or a decision tree.

The next example is similar. We examine the set of VLCs $a_1 = 0$, $a_2 = 10$, $a_3 = 101$, and $a_4 = 111$. Only the code of $a_3$ is different, but a little experimenting shows that this set of VLCs is bad because it is not uniquely decodable (UD). Given the bitstring $0101111 \ldots$, it can be decoded either as $a_1 a_3 a_4 \ldots$ or $a_1 a_2 a_4 \ldots$.

This observation is crucial because it points the way to the construction of large sets of VLCs. The set of codes above is bad because 10, the code of $a_2$, is also the prefix of the code of $a_3$. When the decoder reads $10 \ldots$, it often cannot tell whether this is the code of $a_2$ or the start of the code of $a_3$.

Thus, a useful, practical set of VLCs has to be instantaneous and has to satisfy the following *prefix property*. Once a code $c$ is assigned to a symbol, no other code should start with the bit pattern $c$. Prefix codes are also referred to as prefix-free codes, prefix condition codes, or instantaneous codes.

The following results can be proved: (1) A code is instantaneous if and only if it is a prefix code. (2) The set of UD codes is larger than the set of instantaneous codes (i.e., there are UD codes that are not instantaneous). (3) There is an instantaneous variable-length code with codeword lengths $L_i$ if and only if there is a UD code with these codeword lengths.

The last of these results indicates that we cannot reduce the average word length of a variable-length code by using a UD code rather than an instantaneous code. Thus, there is no loss of compression performance if we restrict our selection of codes to instantaneous codes.

A UD code that consists of $r$ codewords of lengths $l_i$ must satisfy the Kraft inequality (Section 1.5), but this inequality does not require a prefix code. Thus, if a code satisfies the Kraft inequality it is UD, but if it is also a prefix code, then it is instantaneous. This feature of a UD code being also instantaneous, comes for free, because there is no need to add bits to the code and make it longer.

A prefix code (a set of codewords that satisfy the prefix property) is UD. Such a code is also *complete* if adding any codeword to it turns it into a non-UD code. A complete code is the largest UD code, but it also has a downside; it is less robust. If even a single bit is accidentally modified or deleted (or if a bit is somehow added) during storage or transmission, the decoder will lose synchronization and the rest of the transmission will be decoded incorrectly (see the discussion of robust codes in Chapter 3).

While discussing UD and non-UD codes, it is interesting to note that the Morse code is non-UD (because, for example, the code of I is ".." and the code of H is "...."), so Morse had to make it UD by requiring accurate relative timing.

# 1.3 VLCs, Entropy, and Redundancy

Understanding data compression and its codes must start with understanding information, because the former is based on the latter. Hence this short section that introduces a few important concepts from information theory.

Information theory is the creation, in the late 1940s, of Claude Shannon. Shannon tried to develop means for measuring the amount of information stored in a symbol without considering the meaning of the information. He discovered the connection between the logarithm function and information, and showed that the information content (in bits) of a symbol with probability $p$ is $-\log_2 p$. If the base of the logarithm is $e$, then the information is measured in units called nats. If the base is 3, the information units are trits, and if the base is 10, the units are referred to as Hartleys.

Information theory is concerned with the transmission of information from a sender (termed a source), through a communications channel, to a receiver. The sender and receiver can be persons or machines and the receiver may, in turn, act as a sender and send the information it has received to another receiver. The information is sent in units called symbols (normally bits, but in verbal communications the symbols are spoken words) and the set of all possible data symbols is an alphabet.

The most important single factor affecting communications is noise in the communications channel. In verbal communications, this noise is literally noise. When trying to talk in a noisy environment, we may lose part of the discussion. In electronic communications, the channel noise is caused by imperfect hardware and by factors such as sudden lightning, voltage fluctuations—old, high-resistance wires—sudden surge in temperature, and interference from machines that generate strong electromagnetic fields.

The presence of noise implies that special codes should be used to increase the reliability of transmitted information. This is referred to as channel coding or, in everyday language, error-control codes.

The second most important factor affecting communications is sheer volume. Any communications channel has a limited capacity. It can transmit only a limited number of symbols per time unit. An obvious way to increase the amount of data transmitted is to compress it before it is sent (in the source). Methods to compress data are therefore known as source coding or, in everyday language, data compression. The feature that makes it possible to compress data is the fact that individual symbols appear in our data files with different probabilities. One important principle (although not the only one) used to compress data is to assign variable-length codes to the individual data symbols such that short codes are assigned to the common symbols.

Two concepts from information theory, namely entropy and redundancy, are needed in order to fully understand the application of VLCs to data compression.

Roughly speaking, the term "entropy" as defined by Shannon is proportional to the minimum number of yes/no questions needed to reach the answer to some question. Another way of looking at entropy is as a quantity that describes how much information is included in a signal or event. Given a discrete random variable $X$ that can have $n$ values $x_i$ with probabilities $P_i$, the entropy $H(X)$ of $X$ is defined as

$$H(X) = -\sum_{i=1}^{n} P_i \log_2 P_i.$$

A detailed discussion of information theory is outside the scope of this book. Interested readers are referred to the many texts on this subject. Here, we will only show intuitively why the logarithm function plays such an important part in measuring information.

Imagine a source that emits symbols $a_i$ with probabilities $p_i$. We assume that the source is memoryless, i.e., the probability of a symbol being emitted does not depend on what has been emitted in the past. We want to define a function $I(a_i)$ that will measure the amount of information gained when we discover that the source has emitted symbol $a_i$. Function $I$ will also measure our uncertainty as to whether the next symbol will be $a_i$. Alternatively, $I(a_i)$ corresponds to our surprise in finding that the next symbol is $a_i$. Clearly, our surprise at seeing $a_i$ emitted is inversely proportional to the probability $p_i$ (we are surprised when a low-probability symbol is emitted, but not when we notice a high-probability symbol). Thus, it makes sense to require that function $I$ satisfies the following conditions:

1. $I(a_i)$ is a decreasing function of $p_i$, and returns 0 when the probability of a symbol is 1. This reflects our feeling that high-probability events convey less information.

2. $I(a_i a_j) = I(a_i) + I(a_j)$. This is a result of the source being memoryless and the probabilities being independent. Discovering that $a_i$ was immediately followed by $a_j$, provided us with the same information as knowing that $a_i$ and $a_j$ were emitted independently.

Even those with a minimal mathematical background will immediately see that the logarithm function satisfies the two conditions above. This is the first example of the relation between the logarithm function and the quantitative measure of information. The next few paragraphs illustrate other connections between the two.

Consider the case of person A selecting at random an integer $N$ between 1 and 64 and person B having to guess what it is. What is the minimum number of yes/no questions that are needed for B to guess $N$? Those familiar with the technique of binary search know the answer. Using this technique, B should divide the interval 1–64 in two, and should start by asking "is $N$ between 1 and 32?" If the answer is no, then $N$ is in the interval 33 to 64. This interval is then divided by two and B's next question should be "is $N$ between 33 and 48?" This process continues until the interval selected by B shrinks to a single number.

It does not take much to see that exactly six questions are necessary to determine $N$. This is because 6 is the number of times 64 can be divided in half. Mathematically, this is equivalent to writing $6 = \log_2 64$, which is why the logarithm is the mathematical function that quantifies information.

> What we call reality arises in the last analysis from the posing of yes/no questions. All things physical are information-theoretic in origin, and this is a participatory universe.
> —John Wheeler

Another approach to the same problem is to consider a nonnegative integer $N$ and ask how many digits does it take to express it. The answer, of course, depends on $N$. The greater $N$, the more digits are needed. The first 100 nonnegative integers (0 through 99) can be expressed by two decimal digits. The first 1000 such integers can be expressed by three digits. Again it does not take long to see the connection. The number of digits required to represent $N$ equals approximately $\log N$. The base of the logarithm is the same as the base of the digits. For decimal digits, use base 10; for binary digits (bits), use base 2. If we agree that the number of digits it takes to express $N$ is proportional to the information content of $N$, then again the logarithm is the function that gives us a measure of the information. As an aside, the precise length, in bits, of the binary representation of a positive integer $n$ is

$$1 + \lfloor \log_2 n \rfloor \tag{1.1}$$

or, alternatively, $\lceil \log_2(n+1) \rceil$. When $n$ is represented in any other number base $b$, its length is given by the same formula, but with the logarithm in base $b$ instead of 2.

Here is another observation that illuminates the relation between the logarithm and information. A 10-bit string can have $2^{10} = 1024$ values. We say that such a string may contain one of 1024 messages, or that the length of the string is the logarithm of the number of possible messages the string can convey.

The following example sheds more light on the concept of entropy and will prepare us for the definition of redundancy. Given a set of two symbols $a_1$ and $a_2$, with probabilities $P_1$ and $P_2$, respectively, we compute the entropy of the set for various values of the probabilities. Since $P_1 + P_2 = 1$, the entropy of the set is $-P_1 \log_2 P_1 - (1-P_1) \log_2(1-P_1)$ and the results are summarized in Table 1.1.

When $P_1 = P_2$, at least one bit is required to encode each symbol, reflecting the fact that the entropy is at its maximum, the redundancy is zero, and the data cannot be compressed. However, when the probabilities are very different, the minimum number of bits required per symbol drops significantly. We may not be able to conceive a compression method that expresses each symbol in just 0.08 bits, but we know that when $P_1 = 99\%$, such compression is theoretically possible.

| $P_1$ | $P_2$ | Entropy |
|-------|-------|---------|
| 0.99  | 0.01  | 0.08    |
| 0.90  | 0.10  | 0.47    |
| 0.80  | 0.20  | 0.72    |
| 0.70  | 0.30  | 0.88    |
| 0.60  | 0.40  | 0.97    |
| 0.50  | 0.50  | 1.00    |

Table 1.1: Probabilities and Entropies of Two Symbols.

In general, the entropy of a set of $n$ symbols depends on the individual probabilities $P_i$ and is largest when all $n$ probabilities are equal. Data representations often include redundancies and data can be compressed by reducing or eliminating these redundancies. When the entropy is at its maximum, the data has maximum information content and therefore cannot be further compressed. Thus, it makes sense to define redundancy as a quantity that goes down to zero as the entropy reaches its maximum.

> The fundamental problem of communication is that of reproducing at one point either exactly or approximately a message selected at another point.
> —Claude Shannon (1948)

To understand the definition of redundancy, we start with an alphabet of symbols $a_i$, where each symbol appears in the data with probability $P_i$. The data is compressed by replacing each symbol with an $l_i$-bit-long code. The average code length is the sum $\sum P_i l_i$ and the entropy (the smallest number of bits required to represent the symbols) is $\sum [-P_i \log_2 P_i]$. The redundancy $R$ of the set of symbols is defined as the average code length minus the entropy. Thus,

$$R = \sum_i P_i l_i - \sum_i [-P_i \log_2 P_i]. \tag{1.2}$$

The redundancy is zero when the average code length equals the entropy, i.e., when the codes are the shortest and compression has reached its maximum.

Given a set of symbols (an alphabet), we can assign binary codes to the individual symbols. It is easy to assign long codes to symbols, but most practical applications require the shortest possible codes.

Consider the four symbols $a_1$, $a_2$, $a_3$, and $a_4$. If they appear in our data strings with equal probabilities ($= 0.25$), then the entropy of the data is $-4(0.25 \log_2 0.25) = 2$. Two is the smallest number of bits needed, on average, to represent each symbol in this case. We can simply assign our symbols the four 2-bit codes 00, 01, 10, and 11. Since the probabilities are equal, the redundancy is zero and the data cannot be compressed below two bits/symbol.

Next, consider the case where the four symbols occur with different probabilities as shown in Table 1.2, where $a_1$ appears in the data (on average) about half the time, $a_2$ and $a_3$ have equal probabilities, and $a_4$ is rare. In this case, the data has entropy $-(0.49 \log_2 0.49 + 0.25 \log_2 0.25 + 0.25 \log_2 0.25 + 0.01 \log_2 0.01) \sim -(-0.050 - 0.5 - 0.5 - 0.066) = 1.57$. The smallest number of bits needed, on average, to represent each symbol has dropped to 1.57.

| Symbol | Prob. | Code1 | Code2 |
|--------|-------|-------|-------|
| $a_1$ | .49 | 1 | 1 |
| $a_2$ | .25 | 01 | 01 |
| $a_3$ | .25 | 010 | 000 |
| $a_4$ | .01 | 001 | 001 |

Table 1.2: Variable-Length Codes.

If we again assign our symbols the four 2-bit codes 00, 01, 10, and 11, the redundancy would be $R = -1.57 + \log_2 4 = 0.43$. This suggests assigning variable-length codes to the symbols. Code1 of Table 1.2 is designed such that the most common symbol, $a_1$, is assigned the shortest code. When long data strings are transmitted using Code1, the average size (the number of bits per symbol) is $1 \times 0.49 + 2 \times 0.25 + 3 \times 0.25 + 3 \times 0.01 = 1.77$, which is very close to the minimum. The redundancy in this case is $R = 1.77 - 1.57 = 0.2$ bits per symbol. An interesting example is the 20-symbol string $a_1a_3a_2a_1a_3a_3a_4a_2a_1a_1a_2a_2a_1a_1a_3a_1a_1a_2a_3a_1$, where the four symbols occur with approximately the right frequencies. Encoding this string with Code1 yields the 37 bits:

$$1|010|01|1|010|010|001|01|1|1|01|01|1|1|010|1|1|01|010|1$$

(without the vertical bars). Using 37 bits to encode 20 symbols yields an average size of 1.85 bits/symbol, not far from the calculated average size. (The reader should bear in mind that our examples are short. To obtain results close to the best that's theoretically possible, an input stream with at least thousands of symbols is needed.)

However, the conscientious reader may have noticed that Code1 is bad because it is not a prefix code. Code2, in contrast, is a prefix code and can be decoded uniquely. Notice how Code2 was constructed. Once the single bit 1 was assigned as the code of $a_1$, no other codes could start with 1 (they all had to start with 0). Once 01 was assigned as the code of $a_2$, no other codes could start with 01. This is why the codes of $a_3$ and $a_4$ had to start with 00. Naturally, they became 000 and 001.

Designing variable-length codes for data compression must therefore take into account the following two principles: (1) assign short codes to the more frequent symbols and (2) obey the prefix property. Following these principles produces short, unambiguous codes, but not necessarily the best (i.e., shortest) ones. In addition to these principles, an algorithm is needed to generate a set of shortest codes (ones with the minimum average size). The only input to such an algorithm is the frequencies of occurrence (or alternatively the probabilities) of the symbols of the alphabet. The well-known Huffman algorithm (Section 1.13) is such a method. Given a set of symbols whose probabilities of occurrence are known, this algorithm constructs a set of shortest prefix codes for the symbols. Notice that such a set is normally not unique and there may be several sets of codes with the shortest length.

> The beauty of code is much more akin to the elegance, efficiency and clean lines of a spiderweb. It is not the chaotic glory of a waterfall, or the pristine simplicity of a flower. It is an aesthetic of structure, design and order.
>
> —Charles Gordon

Notice that a UD code does not have to be a prefix code. It is possible, for example, to designate the string 111 as a separator (a comma) to separate individual codewords of different lengths, provided that no codeword contains 111. Other examples of a non-prefix, variable-length codes are the $C^3$ code (page 115) and the generalized Fibonacci $C_2$ code (page 118).

# 1.4 Universal Codes

Mathematically, a code is a mapping. It maps source symbols into codewords. Mathematically, a source of messages is a pair $(M, P)$ where $M$ is a (possibly infinite) set of messages and $P$ is a function that assigns a nonzero probability to each message. A message is mapped into a long bitstring whose length depends on the quality of the code and on the probabilities of the individual symbols. The best that can be done is to compress a message to its entropy $H$. A code is universal if it compresses messages to codewords whose average length is bounded by $C1(H + C2)$ where $C1$ and $C2$ are constants greater than or equal to 1, i.e., an average length that is a constant multiple of the entropy plus another constant. A universal code with large constants isn't very useful. A code with $C1 = 1$ is called asymptotically optimal.

A Huffman code often performs better than a universal code, but it can be used only when the probabilities of the symbols are known. In contrast, a universal code can be used in cases where only the ranking of the symbols' probabilities is known. If we know that symbol $a_5$ has the highest probability and $a_8$ has the next largest one, we can assign the shortest codeword to $a_5$ and the next longer codeword to $a_8$. Thus, universal coding amounts to ranking of the source symbols. After ranking, the symbol with index 1 has the largest probability, the symbol with index 2 has the next highest one, and so on. We can therefore ignore the actual symbols and concentrate on their new indexes. We can assign one codeword to index 1, another codeword to index 2, and so on, which is why variable-length codes are often designed to encode integers (Section 2.1). Such a set of variable-length codes can encode any number of integers with codewords that have increasing lengths.

Notice also that a set of universal codes is fixed and so doesn't have to be constructed for each set of source symbols, a feature that simplifies encoding and decoding. However, if we know the probabilities of the individual symbols (the probability distribution of the alphabet of symbols), it becomes possible to tailor the code to the probability, or conversely, to select a known code whose codewords fit the known probability distribution. In all cases, the code selected (the set of codewords) must be uniquely decodable (UD). A non-UD code is ambiguous and therefore useless.

# 1.5 The Kraft–McMillan Inequality

The Kraft–McMillan inequality is concerned with the existence of a uniquely decodable (UD) code. It establishes the relation between such a code and the lengths $L_i$ of its codewords.

One part of this inequality, due to [McMillan 56], states that given a UD variable-length code, with $n$ codewords of lengths $L_i$, the lengths must satisfy the relation

$$\sum_{i=1}^{n} 2^{-L_i} \leq 1. \tag{1.3}$$

The other part, due to [Kraft 49], states the opposite. Given a set of $n$ positive integers $(L_1, L_2, \ldots, L_n)$ that satisfy Equation (1.3), there exists an instantaneous variable-length code such that the $L_i$ are the lengths of its individual codewords.

Together, both parts say that there is an instantaneous variable-length code with codeword lengths $L_i$ if and only if there is a UD code with these codeword lengths. The two parts do not say that a variable-length code is instantaneous or UD if and only if the codeword lengths satisfy Equation (1.3). In fact, it is easy to check the three individual code lengths of the code $(0, 01, 011)$ and verify that $2^{-1} + 2^{-2} + 2^{-3} = 7/8$. This code satisfies the Kraft–McMillan inequality and yet it is not instantaneous, because it is not a prefix code. Similarly, the code $(0, 01, 001)$ also satisfies Equation (1.3), but is not UD. A few more comments on this inequality are in order:

- If a set of lengths $L_i$ satisfies Equation (1.3), then there exist instantaneous and UD variable-length codes with these lengths. For example $(0, 10, 110)$.

- A UD code is not always instantaneous, but there exists an instantaneous code with the same codeword lengths. For example, code $(0, 01, 11)$ is UD but not instantaneous, while code $(0, 10, 11)$ is instantaneous and has the same lengths.

- The sum of Equation (1.3) corresponds to the part of the complete code tree that has been used for codeword selection. This is why the sum has to be less than or equal to 1. This intuitive explanation of the Kraft–McMillan relation is explained in the next paragraph.

We can gain a deeper understanding of this useful and important inequality by constructing the following simple prefix code. Given five symbols $a_i$, suppose that we decide to assign 0 as the code of $a_1$. Now all the other codes have to start with 1. We therefore assign 10, 110, 1110, and 1111 as the codewords of the four remaining symbols. The lengths of the five codewords are 1, 2, 3, 4, and 4, and it is easy to see that the sum

$$2^{-1} + 2^{-2} + 2^{-3} + 2^{-4} + 2^{-4} = \frac{1}{2} + \frac{1}{4} + \frac{1}{8} + \frac{2}{16} = 1$$

satisfies the Kraft–McMillan inequality. We now consider the possibility of constructing a similar code with lengths 1, 2, 3, 3, and 4. The Kraft–McMillan inequality tells us that this is impossible, because the sum

$$2^{-1} + 2^{-2} + 2^{-3} + 2^{-3} + 2^{-4} = \frac{1}{2} + \frac{1}{4} + \frac{2}{8} + \frac{1}{16}$$

is greater than 1, and this is easy to understand when we consider the code tree. Starting with a complete binary tree of height 4, such as the tree of Figure 3.17, it is obvious that once 0 was assigned as a codeword, we have "used" one half of the tree and all future codes would have to be selected from the other half of the tree. Once 10 was assigned, we were left with only 1/4 of the tree. Once 110 was assigned as a codeword, only 1/8 of the tree remained available for the selection of future codes. Once 1110 has been assigned, only 1/16 of the tree was left, and that was enough to select and assign code 1111. However, once we select and assign codes of lengths 1, 2, 3, and 3, we have exhausted the entire tree and there is nothing left to select the last (4-bit) code from.

The Kraft–McMillan inequality can be related to the entropy by observing that the lengths $L_i$ can always be written as $L_i = -\log_2 P_i + E_i$, where $E_i$ is simply the amount by which $L_i$ is greater than the entropy (the extra length of code $i$).

This implies that

$$2^{-L_i} = 2^{(\log_2 P_i - E_i)} = 2^{\log_2 P_i}/2^{E_i} = P_i/2^{E_i}.$$

In the special case where all the extra lengths are the same ($E_i = E$), the Kraft–McMillan inequality says that

$$1 \geq \sum_{i=1}^{n} P_i/2^E = \frac{\sum_{i=1}^{n} P_i}{2^E} = \frac{1}{2^E} \implies 2^E \geq 1 \implies E \geq 0.$$

An unambiguous code has nonnegative extra length, meaning its length is greater than or equal to the length determined by its entropy.

Here is a simple example of the use of this inequality. Consider the simple case of $n$ equal-length binary codewords. The size of each codeword is $L_i = \log_2 n$, and the Kraft–McMillan sum is

$$\sum_{1}^{n} 2^{-L_i} = \sum_{1}^{n} 2^{-\log_2 n} = \sum_{1}^{n} \frac{1}{n} = 1.$$

The inequality is satisfied, so such a code is UD.

A more interesting example is the case of $n$ symbols where the first one is compressed and the second one is expanded. We set $L_1 = \log_2 n - a$, $L_2 = \log_2 n + e$, and $L_3 = L_4 = \cdots = L_n = \log_2 n$, where $a$ and $e$ are positive. We show that $e > a$, which means that compressing a symbol by a factor $a$ requires expanding another symbol by a larger factor. We can benefit from this only if the probability of the compressed symbol is greater than that of the expanded symbol.

$$\sum_{1}^{n} 2^{-L_i} = 2^{-L_1} + 2^{-L_2} + \sum_{3}^{n} 2^{-\log_2 n}$$

$$= 2^{-\log_2 n + a} + 2^{-\log_2 n - e} + \sum_{1}^{n} 2^{-\log_2 n} - 2 \times 2^{-\log_2 n}$$

$$= \frac{2^a}{n} + \frac{2^{-e}}{n} + 1 - \frac{2}{n}.$$

The Kraft–McMillan inequality requires that

$$\frac{2^a}{n} + \frac{2^{-e}}{n} \leq 1, \quad \frac{2}{n} \leq 1, \quad \frac{2^a}{n} + \frac{2^{-e}}{n} - \frac{2}{n} < 0,$$

or $2^{-e} \leq 2 - 2^a$, implying $-e \leq \log_2(2 - 2^a)$, or $e \geq -\log_2(2 - 2^a)$.

The inequality above implies $a \leq 1$ (otherwise, $2 - 2^a$ is negative) but $a$ is also positive (since we assumed compression of symbol 1). The possible range of values of $a$ is therefore $(0, 1]$, and in this range $e$ is greater than $a$, proving the statement above. (It is easy to see that $a = 1 \rightarrow e \geq -\log_2 0 = \infty$, and $a = 0.1 \rightarrow e \geq -\log_2(2 - 2^{0.1}) \approx 0.10745$.)

It can be shown that this is just a special case of a general result that says, given an alphabet of $n$ symbols, if we compress some of them by a certain factor, then the others must be expanded by a greater factor.

> One of my most productive days was throwing away 1000 lines of code.
> —Kenneth Thompson

## 1.6 Tunstall Code

The main advantage of variable-length codes is their variable lengths. Some codes are short, others are long, and a clever assignment of codes to symbols can produce compression. On the downside, variable-length codes are difficult to work with. The encoder has to construct each code from individual bits and pieces, has to accumulate and append several such codes in a short buffer, wait until $n$ bytes of the buffer are full of code bits (where $n$ must be at least 1), write the $n$ bytes onto the output, shift the buffer $n$ bytes, and keep track of the location of the last bit placed in the buffer. The decoder has to go through the reverse process. It is definitely easier to deal with fixed-size codes, and the Tunstall codes described here are an example of how such codes can be designed. The idea is to construct a set of fixed-size codes, each encoding a variable-length string of input symbols. As a result, these codes are also known as variable-to-fixed (or variable-to-block) codes, in contrast to the variable-length codes which are also referred to as fixed-to-variable.

Imagine an alphabet that consists of two symbols $A$ and $B$ where $A$ is more common. Given a typical string from this alphabet, we expect substrings of the form $AA$, $AAA$, $AB$, $AAB$, and $B$, but rarely strings of the form $BB$. We can therefore assign fixed-size codes to the following five substrings as follows. $AA = 000$, $AAA = 001$, $AB = 010$, $ABA = 011$, and $B = 100$. A rare occurrence of two consecutive $B$s will be encoded by 100100, but most occurrences of $B$ will be preceded by an $A$ and will be coded by 010, 011, or 100.

This example is both bad and inefficient. It is bad, because $AAABAAB$ can be encoded either as the four codes $AAA$, $B$, $AA$, $B$ or as the three codes $AA$, $ABA$, $AB$; encoding is not unique and may require several passes to determine the shortest code. This happens because our five substrings don't satisfy the prefix property. This example is inefficient because only five of the eight possible 3-bit codes are used. An $n$-bit Tunstall code should use all $2^n$ codes. Another point is that our codes were selected without considering the relative frequencies of the two symbols, and as a result we cannot be certain that this is the best code for our alphabet.

Thus, an algorithm is needed to construct the best $n$-bit Tunstall code for a given alphabet of $N$ symbols, and such an algorithm is given in [Tunstall 67]. Given an alphabet of $N$ symbols, we start with a code table that consists of the symbols. We then iterate as long as the size of the code table is less than or equal to $2^n$ (the number of $n$-bit codes). Each iteration performs the following steps:

■   Select the symbol with largest probability in the table. Call it $S$.

■   Remove $S$ and include the $N$ substrings $Sx$ where $x$ goes over all the $N$ symbols. This step increases the table size by $N - 1$ symbols (some of them may be substrings). Thus, after iteration $k$, the table size will be $N + k(N - 1)$ elements.

■   If $N + (k + 1)(N - 1) \leq 2^n$, perform another iteration (iteration $k + 1$).

It is easy to see that the elements (symbols and substrings) of the table satisfy the prefix property and thus ensure unique encodability. If the first iteration adds element $AB$ to the table, it must have removed element $A$. Thus, $A$, the prefix of $AB$, is not a code. If the next iteration creates element $ABR$, then it has removed element $AB$, so $AB$ is not a prefix of $ABR$. This construction also minimizes the average number of bits per alphabet symbol because of the requirement that each iteration select the element (or an element) of maximum probability. This requirement is similar to the way a Huffman code is constructed, and we illustrate it by an example.

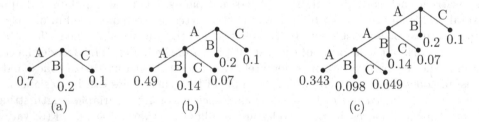

Figure 1.3: Tunstall Code Example.

Given an alphabet with the three symbols $A$, $B$, and $C$ ($N = 3$), with probabilities 0.7, 0.2, and 0.1, respectively, we decide to construct a set of 3-bit Tunstall codes (thus, $n = 3$). We start our code table as a tree with a root and three children (Figure 1.3a). In the first iteration, we select $A$ and turn it into the root of a subtree with children $AA$, $AB$, and $AC$ with probabilities 0.49, 0.14, and 0.07, respectively (Figure 1.3b). The largest probability in the tree is that of node $AA$, so the second iteration converts it to the root of a subtree with nodes $AAA$, $AAB$, and $AAC$ with probabilities 0.343, 0.098, and 0.049, respectively (Figure 1.3c). After each iteration we count the number of leaves of the tree and compare it to $2^3 = 8$. After the second iteration there are seven leaves in the tree, so the loop stops. Seven 3-bit codes are arbitrarily assigned to elements $AAA$, $AAB$, $AAC$, $AB$, $AC$, $B$, and $C$. The eighth available code should be assigned to a substring that has the highest probability and also satisfies the prefix property.

The average bit length of this code is easily computed as

$$\frac{3}{3(0.343 + 0.098 + 0.049) + 2(0.14 + 0.07) + 0.2 + 0.1} = 1.37 \text{ bits/symbol.}$$

In general, let $p_i$ and $l_i$ be the probability and length of tree node $i$. If there are $m$ nodes in the tree, the average bit length of the Tunstall code is $n/\sum_{i=1}^{m} p_i l_i$. The entropy of our alphabet is $-(0.7 \times \log_2 0.7 + 0.2 \times \log_2 0.2 + 0.1 \times \log_2 0.1) = 1.156$, so the Tunstall codes do not provide the best compression.

The tree of Figure 1.3 is referred to as a parse tree, not a code tree. It is complete in the sense that every interior node has $N$ children. Notice that the total number of nodes of this tree is $3 \times 2 + 1$ and in general $a(N - 1) + 1$. A parse tree defines a set of substrings over the alphabet (seven substrings in our example) such that any string of symbols from the alphabet can be broken up (subdivided) into these substrings (except that the last part may be only a prefix of such a substring) in one way only. The subdivision is unique because the set of substrings defined by the parse tree is proper, i.e., no substring is a prefix of another substring.

An important property of the Tunstall codes is their reliability. If one bit becomes corrupt, only one code will get bad. Normally, variable-length codes are not robust. One bad bit may corrupt the decoding of the remainder of a long sequence of such codes. It is possible to incorporate error-control codes in a string of variable-length codes, but this increases its size and reduces compression.

Section 1.13.1 illustrates how a combination of the Tunstall algorithm with Huffman coding can improve compression in a two-step, dual tree process.

A major downside of the Tunstall code is that both encoder and decoder have to store the complete code (the set of substrings of the parse tree).

> There are 10 types of people in this world: those who understand binary and those who don't.
>
> —Author unknown

# 1.7 Schalkwijk's Coding

One of the many contributions of Isaac Newton to mathematics is the well-known binomial theorem. It states

$$(a + b)^n = \sum_{i=0}^{n} \binom{n}{i} a^i b^{n-i},$$

where the term

$$\binom{n}{i} = \frac{n!}{i!(n - i)!}$$

is pronounced "$n$ over $i$" and is referred to as a binomial coefficient.

```
                                1
                             1     1
                          1     2     1
                       1     3     3     1
                    1     4     6     4     1
                 1     5    10    10     5     1
              1     6    15    20    15     6     1
           1     7    21    35    35    21     7     1
        1     8    28    56    70    56    28     8     1
     1     9    36    84   126   126    84    36     9     1
  1    10    45   120   210   252   210   120    45    10     1
1    11    55   165   330   462   462   330   165    55    11     1
1   12    66   220   495   792   924   792   495   220    66    12    1
1   13    78   286   715  1287  1716  1716  1287   715   286    78    13    1
1  14    91   364  1001  2002  3003  3432  3003  2002  1001   364    91   14   1
1  15   105   455  1365  3003  5005  6435  6435  5005  3003  1365   455  105   15   1
1  16   120   560  1820  4368  8008 11440 12870 11440  8008  4368  1820  560  120  16   1
```

Figure 1.4: Pascal Triangle.

Blaise Pascal (Section 2.23), a contemporary of Newton, discovered an elegant way to compute these coefficients without the lengthy calculations of factorials, multiplications, and divisions. He conceived the famous triangle that now bears his name (Figure 1.4) and showed that the general element of this triangle is a binomial coefficient.

> Quand on voit le style naturel, on est tout ètonnè et ravi, car on s'attendait de voir un auteur, et on trouve un homme. (When we see a natural style, we are quite surprised and delighted, for we expected to see an author and we find a man.)
>
> —Blaise Pascal, *Pensèes* (1670)

The Pascal triangle is an infinite triangular matrix that's constructed from the edges inwards. First fill the left and right edges with 1's, then compute each interior element as the sum of the two elements directly above it. The construction is simple and it is trivial to derive an explicit expression for the general element of the triangle and show that it is a binomial coefficient. Number the rows from 0 starting at the top, and number the columns from 0 starting on the left. A general element is denoted by $\binom{i}{j}$. Now observe that the top two rows (corresponding to $i = 0$ and $i = 1$) consist of 1's and that every other row can be obtained as the sum of its predecessor and a shifted version of its predecessor. For example,

$$
\begin{array}{r}
1\ 3\ 3\ 1\phantom{\ 1} \\
+\quad 1\ 3\ 3\ 1 \\
\hline
1\ 4\ 6\ 4\ 1
\end{array}
$$

This shows that the elements of the triangle satisfy

$$\binom{i}{0} = \binom{i}{i} = 1, \qquad i = 0, 1, \ldots,$$

$$\binom{i}{j} = \binom{i-1}{j-1} + \binom{i-1}{j}, \qquad i = 2, 3, \ldots, \quad j = 1, \ldots, (i-1).$$

From this, it is easy to obtain the explicit expression

$$\binom{i}{j} = \binom{i-1}{j-1} + \binom{i-1}{j}$$

$$= \frac{(i-1)!}{(j-1)!(i-j)!} + \frac{(i-1)!}{j!(i-1-j)!}$$

$$= \frac{j(i-1)!}{j!(i-j)!} + \frac{(i-j)(i-1)!}{j!(i-j)!}$$

$$= \frac{i!}{j!(i-j)!}.$$

And this is Newton's binomial coefficient $\binom{i}{j}$.

---

The Pascal triangle has many interesting and unexpected properties, some of which are listed here.

■  The sum of the elements of row $i$ is $2^i$.

■  If the second element of a row is a prime number, all the elements of the row (except the 1's) are divisible by it. For example, the elements 7, 21, and 35 of row 7 are divisible by 7.

■  Select any diagonal and any number of consecutive elements on it. Their sum will equal the number on the row below the end of the selection and off the selected diagonal. For example, $1 + 6 + 21 + 56 = 84$.

■  Select row 7, convert its elements 1, 7, 21, 35, 35, 21, 7, and 1 to the single number 19487171 by concatenating the elements, except that a multidigit element is first carried over, such that 1, 7, 21, 35,... become $1(7+2)(1+3)(5+3)\ldots = 1948\ldots$. This number equals $11^7$ and this magic-11 property holds for any row.

■  The third column from the right consists of the triangular numbers 1, 3, 6, 10,....

■  Select all the odd numbers on the triangle and fill them with black. The result is the Sierpinski triangle (a well-known fractal).

Other unusual properties can be found in the vast literature that exists on the Pascal triangle. The following is a quotation from Donald Knuth:

"There are so many relations in Pascal's triangle, that when someone finds a new identity, there aren't many people who get excited about it anymore, except the discoverer."

---

The Pascal triangle is the basis of the unusual coding scheme described in [Schalkwijk 72]. This method starts by considering all the finite bit strings of length $n$ that have exactly $w$ 1's. The set of these strings is denoted by $T(n, w)$. If a string $t$ consists of bits $t_1$ through $t_n$, then we define weights $w_1$ through $w_n$ as the partial sums

$$w_k = \sum_{i=k}^{n} t_i.$$

Thus, if $t = 010100$, then $w_1 = w_2 = 2$, $w_3 = w_4 = 1$, and $w_5 = w_6 = 0$. Notice that $w_1$ always equals $w$.

We now define, somewhat arbitrarily, a ranking $i(t)$ on the $\binom{n}{w}$ strings in set $T(n, w)$ by

$$i(t) = \sum_{k=1}^{n} t_k \binom{n-k}{w_k}.$$

(The binomial coefficient $\binom{i}{j}$ is defined only for $i \geq j$, so we set it to 0 when $i < n$.) The rank of $t = 010100$ becomes

$$0 + 1\binom{6-2}{2} + 0 + 1\binom{6-4}{1} + 0 + 0 = 6 + 2 = 8.$$

It can be shown that the rank of a string $t$ in set $T(n, w)$ is between 0 and $\binom{n}{w} - 1$.

The following table lists the ranking for the $\binom{6}{2} = 15$ strings of set $T(6, 2)$.

| 0 | 000011 | 3 | 001001 | 6 | 010001 | 9 | 011000 | 12 | 100100 |
|---|--------|---|--------|---|--------|---|--------|----|--------|
| 1 | 000101 | 4 | 001010 | 7 | 010010 | 10 | 100001 | 13 | 101000 |
| 2 | 000110 | 5 | 001100 | 8 | 010100 | 11 | 100010 | 14 | 110000 |

The first version of the Schalkwijk coding algorithm is not general. It is restricted to data symbols that are elements $t$ of $T(n, w)$. We assume that both encoder and decoder know the values of $n$ and $w$. The method employs the Pascal triangle to determine the rank $i(t)$ of each string $t$, and this rank becomes the code of $t$. The maximum value of the rank is $\binom{n}{w} - 1$, so it can be expressed in $\lceil \log_2 \binom{n}{w} \rceil$ bits. Thus, this method compresses each $n$-bit string (of which $w$ bits are 1's) to $\lceil \log_2 \binom{n}{w} \rceil$ bits.

Consider a source of bits that emits a 0 with probability $q$ and a 1 with probability $p = 1 - q$. The entropy of this source is $H(p) = -p \log_2 p - (1 - p) \log_2 (1 - p)$. In our strings, $p = w/n$, so the compression performance of this method is measured by the ratio

$$\frac{\lceil \log_2 \binom{n}{w} \rceil}{n}$$

and it can be shown that this ratio approaches $H(w/n)$ when $n$ becomes very large. (The proof employs the famous Stirling formula $n! \approx \sqrt{2\pi n}\, n^n e^{-n}$.)

Figure 1.5a illustrates the operation of both encoder and decoder. Both know the values of $n$ and $w$ and they construct in the Pascal triangle a coordinate system tilted as shown in the figure and with its origin at element $w$ of row $n$ of the triangle.

As an example, suppose that $(n, w) = (6, 2)$. This puts the origin at the element 15 as shown in part (a) of the figure. The encoder starts at the origin, reads bits from the input string, and moves one step in the $x$ direction for each 0 read and one step in the $y$ direction for each 1 read. In addition, before moving in the $y$ direction, the encoder saves the next triangle element in the $x$ direction (the one it will not go to). Thus, given the string 010100, the encoder starts at the origin (15), moves to 10, 4, 3, 1, 1, and 1, while saving the values 6 (before it moves from 10 to 4) and 2 (before it moves from 3 to 1). The sum $6 + 2 = 8 = 1000_2$ is the 4-bit rank of the input string and it becomes the encoder's output.

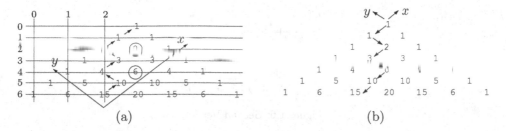

Figure 1.5: First Version.

The decoder also knows the values of $n$ and $w$, so it constructs the same coordinate system in the triangle and starts at the origin. Given the 4-bit input $1000_2 = 8$, the decoder compares it to the next $x$ value 10, and finds that $8 < 10$. It therefore moves in the $x$ direction, to 10, and emits a 0. The input 8 is compared to the next $x$ value 6, but it is not less than 6. The decoder responds by subtracting $8 - 6 = 2$, moving in the $y$ direction, to 4, and emitting a 1. The current input, 2, is compared to the next $x$ value 3, and is found to be smaller. The decoder therefore moves in the $x$ direction, to 3, and emits a 0. When the input 2 is compared to the next $x$ value 2, it is not smaller, so the decoder: (1) subtracts $2 - 2 = 0$, (2) moves in the $y$ direction to 1, and (3) emits a 1. The decoder's input is now 0, so the decoder finds it smaller than the values on the $x$ axis. It therefore keeps moving in the $x$ direction, emitting two more zeros until it reaches the top of the triangle.

A similar variant is shown in Figure 1.5b. The encoder always starts at the apex of the triangle, moves in the $-x$ direction for each 0 and in the $-y$ direction for each 1, where it also records the value of the next element in the $-x$ direction. Thus, the two steps in the $-y$ direction in the figure record the values 1 and 3, whose sum 4 becomes the encoded value of string 010100. The decoder starts at 15 and proceeds in the opposite direction toward the apex. It is not hard to see that it ends up decoding the string 001010, which is why the decoder's output in this variant has to be reversed before it is used.

This version of Schalkwijk coding is restricted to certain bit strings, and is also block-to-block coding. Each block of $n$ bits is replaced by a block of $\lceil \log_2 \binom{n}{w} \rceil$ bits. The next version is similar, but is variable-to-block coding. We again assume a source of bits that emits a 0 with probability $q$ and a 1 with probability $p = 1 - q$. A string of $n$ bits from this source may often have close to $pn$ 1's and $qn$ zeros, but may sometimes have different numbers of zeros and 1's. We select a convenient value for $n$, a value that is as large as possible and where both $pn$ and $qn$ are integers or very close to integers. If $p = 1/3$, for example, then $n = 12$ may make sense, because it results in $np = 4$ and $nq = 8$. We again employ the Pascal triangle and take a rectangular block of $(pn + 1)$ rows and $(qn + 1)$ columns such that the top of the triangle will be at the top-right corner of the rectangle (Figure 1.6).

As before, we start at the bottom-left corner of the array and read bits from the source. For each 0 we move a step in the $x$ direction and for each 1 we move in the $y$ direction. If the next $n$ bits have exactly $pn$ 1's, we will end up at point "A," the top-right corner of the array, and encode $n$ bits as before. If the first $n$ bits happen to

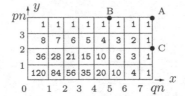

Figure 1.6: Second Version.

have more than $pn$ 1's, then the top of the array will be reached (after we have read $np$ 1's) early, say at point "B," before we have read $n$ bits. We cannot read any more source bits, because any 1 would take us outside the array, so we append several dummy zeros to what we have read, to end up with $n$ bits (of which $np$ are 1's). This is encoded as before. Notice that the decoder can mimic this operation. It operates as before, but stops decoding when it reaches the top boundary. If the first $n$ bits happen to have many zeros, the encoder will end up at the right boundary of the array, say, at point "C," after it has read $qn$ zeros but before it has read $n$ bits. In such a case, the encoder appends several 1's, to end up with exactly $n$ bits (of which precisely $pn$ are 1's), and encodes as before. The decoder can mimic this operation by simply stopping when it reaches the right boundary.

Any string that has too many or too few 1's degrades the compression, because it encodes fewer than $n$ bits in the same $\lceil \log_2 \binom{n}{pn} \rceil$ bits. Thus, the method may not be very effective, but it is an example of a variable-to-block encoding.

The developer of this method points out that the method can be modified to employ the Pascal triangle for block-to-variable coding. The value of $n$ is determined and it remains fixed. Blocks of $n$ bits are encoded and each block is preceded by the number of 1's it contains. If the block contains $w$ 1's, it is encoded by the appropriate part of the Pascal triangle. Thus, each block of $n$ bits may be encoded by a different part of the triangle, thereby producing a different-length code. The decoder can still work in lockstep with the encoder, because it first reads the number $w$ of 1's in a block. Knowing $n$ and $w$ tells it what part of the triangle to use and how many bits of encoding to read. It has been pointed out that this variant is similar to the method proposed by [Lynch 66] and [Davisson 66]. This variant has also been extended by [Lawrence 77], whose block-to-variable coding scheme is based on a Pascal triangle where the boundary points are defined in a special way, based on the choice of a parameter $S$.

# 1.8 Tjalkens–Willems V-to-B Coding

The little-known variable-to-block coding scheme presented in this section is due to [Tjalkens and Willems 92] and is an extension of earlier work described in [Lawrence 77]. Like the Schalkwijk's codes of Section 1.7 and the Lawrence algorithm, this scheme employs the useful properties of the Pascal triangle. The method is based on the choice of a positive integer parameter $C$. Once a value for $C$ has been selected, the authors show how to construct a set $L$ of $M$ variable-length bitstrings that satisfy the following:

1. Set $L$ is complete. Given any infinite bitstring (in practice, a string of $M$ or more bits), $L$ contains a prefix of the string.

2. Out $L$ is proper. No segment in the set is a prefix of another segment.

Once $L$ has been constructed, it is kept in lexicographically sorted order, so each string in $L$ has an index between 0 and $M - 1$. The input to be encoded is a long bitstring. It is broken up by the encoder into segments of various lengths that are members of $L$. Each segment is encoded by replacing it with its index in $L$. Note that the index is a $(\log_2 M)$-bit number. Thus, if $M = 256$, each segment is encoded in eight bits. The main task is to construct set $L$ in such a way that the encoder will be able to read the input bit by bit, stop when it has read a bit pattern that is a string in $L$, and determine the code of the string (its index in $L$). The theory behind this method is complex, so only the individual steps and tests are summarized here.

Given a string $s$ of $a$ zeros and $b$ 1's, we define the function

$$Q(s) = (a + b + 1)\binom{a + b}{b}.$$

(The authors show that $1/Q$ is the probability of string $s$.) We denote by $s_{-1}$ the string $s$ without its last (rightmost) bit. String $s$ is included in set $L$ if it satisfies

$$Q(s_{-1}) < C \le Q(s). \tag{1.4}$$

The authors selected $C = 82$ (because this results in the convenient size $M = 256$). Once $C$ is known, it is easy to decide whether a given string $s$ with $a$ zeros and $b$ 1's is a member of set $L$ (i.e., whether $s$ satisfies Equation (1.4)). If $s$ is in $L$, then point $(a, b)$ in the Pascal triangle (i.e., element $b$ of row $a$, where row and column numbering starts at 0) is considered a boundary point. Figure 1.7a shows the boundary points (underlined) for $C = 82$.

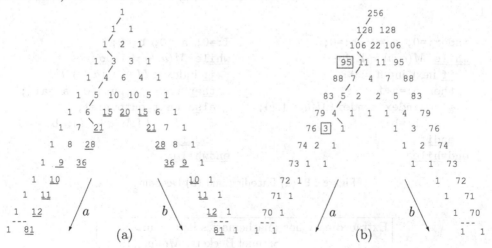

Figure 1.7: (a) Boundary Points. (b) Coding Array.

The inner parts of the triangle are not used in this method and can be removed. Also, The lowest boundary points are located on row 81 and lower parts of the triangle are not used. If string $s$ is in set $L$, then we can start at the apex of the triangle, move in the $a$ direction for each 0 and in the $b$ direction for each 1 in $s$, and end up at a boundary point. The figure illustrates this walk for the string 0010001, where the boundary point reached is 21.

Setting up the partial Pascal triangle of Figure 1.7a is just the first step. The second step is to convert this triangle to a coding triangle $M(a, b)$ of the same size, where each walk for a string $s$ can be used to determine the index of $s$ in set $L$ and thus its code. The authors show that element $M(a, b)$ of this triangle must equal the number of distinct ways to reach a boundary point after processing $a$ zeros and $b$ 1's (i.e., after moving $a$ steps in the $a$ direction and $b$ steps in the $b$ direction). Triangle $M(a, b)$ is constructed according to

$$M(a,b) = \begin{cases} 1, & \text{if } (a,b) \text{ is a boundary point,} \\ M(a+1,b) + M(a,b+1), & \text{otherwise.} \end{cases}$$

The coding array for $C = 82$ (constructed from the bottom up) is shown in Figure 1.7b. Notice that its apex, $M(0,0)$, equals the total number of strings in $L$. Once this triangle is available, both encoding and decoding are simple and are listed in Figure 1.8a,b. The former inputs individual bits and moves in $M(a, b)$ in the $a$ or $b$ directions according to the inputs. The end of the current input string is signalled when a node with a 1 is reached in the coding triangle. For each move in the $b$ direction, the next element in the $a$ direction (the one that will not be reached) is added to the index. At the end, the index is the code of the current string. Figure 1.7b shows the moves for 0010001 and how the nodes 95 and 3 are selected and added to become code 98. The decoder starts with the code of a string in variable index. It compares index to the sum $(I + M(a + 1, b))$ and moves in the $a$ or $b$ directions according to the result, generating one output bit as it moves. Decoding is complete when the decoder reaches a node with a 1.

```
index:=0; a:=0; b:=0;                I:=0; a:=0; b:=0;
while M(a,b) ≠ 1 do                  while M(a,b) ≠ 1 do
   if next_input = 0                    if index < (I + M(a+1,b))
   then a:=a+1                          then next_output:=0; a:=a+1;
   else index:=index+M(a+1,b);          else next_output:=1;
        b:=b+1                               I := I+M(a+1,b); b:=b+1
   endif                               endif
endwhile                             endwhile
```

Figure 1.8: (a) Encoding and (b) Decoding.

Extraordinary how mathematics help you....
—Samuel Beckett, *Molloy* (1951)

# 1.9 Phased-In Codes

Many of the prefix codes described here were developed for the compression of specific types of data. These codes normally range from very short to indefinitely long, and they are suitable for the compression of data where individual symbols have small and large probabilities. Data where symbols have equal probabilities cannot be compressed by VLCs and may be assigned fixed-length codes. The codes of this section (also called phased-in binary codes, see Appendix A-2 in [Bell et al. 90]) constitute a compromise. A set of phased-in codes consists of codes of two lengths and may contribute a little to the compression of data where symbols have equal or almost equal probabilities.

Here is an example for the case $n = 24$. Given a set of 24 symbols $a_0$ through $a_{23}$, we first determine that the largest power of 2 in the interval $[0, 23]$ is 16. The first $2^4 = 16$ symbols $a_i$ are assigned the codes $i + 16$. These codes are the 5-bit numbers $16 = 10000_2$ through $31 = 11111_2$. The remaining symbols $a_{16}$ through $a_{23}$ are assigned codes $i - 16$, resulting in the 4-bit numbers $0 = 0000_2$ through $7 = 0111_2$. The final result is a set of the sixteen 5-bit codes 10000 through 11111, followed by the eight 4-bit codes 0000 through 0111.

Decoding is straightforward. First read four bits into $T$. If $T \leq 7$, then the code is the 4-bit $T$; otherwise, read the next bit $u$ and compute the 5-bit code $2T + u$.

In general, we assume an alphabet that consists of the $n$ symbols $a_0, a_1, \ldots, a_{n-1}$. We select the integer $m$ that satisfies $2^m \leq n < 2^{m+1}$. The first $2^m$ symbols $a_0$ through $a_{2^m-1}$ are encoded as the $(m+1)$-bit numbers $i + 2^m$. This results in codes $2^m$ through $2^{m+1} - 1$. The remaining $n - 2^m$ symbols $a_{2^m}$ through $a_{n-1}$ are encoded as the $m$-bit numbers $i - 2^m$. This results in codes 0 through $n - 2^m$.

To decode, read the first $m$ bits into $T$. If $T \leq n - 2^m$, then the code is the $m$-bit $T$; otherwise, read the next bit $u$ and compute the $(m+1)$-bit code $2T + u$.

The phased-in codes are closely related to the minimal binary code of Section 2.12.

The efficiency of phased-in codes is easy to estimate. The first $2^m$ symbols are encoded in $m + 1$ bits each and the remaining $n - 2^m$ symbols are encoded in $m$ bits each. The average number of bits for each of the $n$ symbols is therefore $[2^m(m + 1) + (n - 2^m)m]/n = (2^m/n) + m$. Fixed-length (block) codes for the $n$ symbols are $m + 1$ bits each, so the quantity $[(2^m/n) + m]/(m + 1)$ is a good measure of the efficiency of this code. For $n = 2^m$, this measure equals 1, while for other values of $n$ it is less than 1, as illustrated in Figure 1.9.

One application of these codes is as pointers to a table. Given a table of 1000 entries, pointers to the table are in the interval $[0, 999]$ and are normally ten bits long, but not all the 1024 10-bit values are needed. If phased-in codes are used to encode the pointers, they become either 10 or nine bits each, resulting in a small compression of the set of pointers. It is obvious that $2^9 \leq 1000 < 2^{10}$, so $m = 9$, resulting in the 512 10-bit codes $0 + 2^9 = 512$ through $511 + 2^9 = 1023$ and the 488 9-bit codes $512 - 2^9 = 0$ to $999 - 2^9 = 487$. The average length of a pointer is now $(512 \times 10 + 488 \times 9)/1000 = 9.512$ bits.

The application of phased-in codes in this case is effective because the number of data items is close to $2^{m+1}$. In cases where the table size is close to $2^m$, however, the phased-in codes are not that efficient. A simple example is a table with $2^9 + 1 = 513$ entries. The value of $m$ is again 9, and the first 512 phased-in codes are the 10-bit

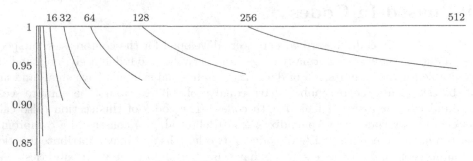

Figure 1.9: Efficiency of Phased-In Codes.

```
g=Table[Plot[((2^m/n)+m)/(m+1), {n,2^m,2^(m+1)-0.99}], {m,0,8}];
Show[g]
```

Code for Figure 1.9.

numbers $0 + 2^9 = 512$ through $511 + 2^9 = 1023$. The 513th code is the 9-bit number $512 - 2^9 = 0$. The average code size is now $(512 \times 10 + 1 \times 9)/513 \approx 9.99$ bits.

See [seul.org 06] for a *Mathematica* notebook to construct phased-in codes.

It is also possible to construct suffix phased-in codes, where the leftmost bit of some codes is removed if it is a 0 and if its removal does not create any ambiguity. Table 1.10 (where the removed bits are in italics) illustrates an example for the first 24 nonnegative integers. The fixed-sized representation of these integers requires five bits, but each of the eight integers 8 through 15 can be represented by only four bits because 5-bit codes can represent 32 symbols and we have only 24 symbols. A simple check verifies that, for example, coding the integer 8 as 1000 instead of 01000 does not introduce any ambiguity, because none of the other 23 codes ends with 1000. One-third of the codewords in this example are one bit shorter, but if we consider only the 17 integers from 0 to 16, about half will require four bits instead of five. The efficiency of this code depends on where $n$ (the number of symbols) is located in the interval $[2^m, 2^{m+1} - 1)$.

| | | | | | | | |
|---|---|---|---|---|---|---|---|
| 00000 | 00001 | 00010 | 00011 | 00100 | 00101 | 00110 | 00111 |
| *0*1000 | *0*1001 | *0*1010 | *0*1011 | *0*1100 | *0*1101 | *0*1110 | *0*1111 |
| 10000 | 10001 | 10010 | 10011 | 10100 | 10101 | 10110 | 10111 |

Table 1.10: Suffix Phased-In Codes.

The suffix phased-in codes are suffix codes (if $c$ has been selected as a codeword, no other codeword will end with $c$). Suffix codes can be considered the complements of prefix codes and are also mentioned in Section 3.5.

# 1.10 Redundancy Feedback (RF) Coding

The interesting and original method of redundancy feedback (RF) coding is the brain-child of Eduardo Enrique González Rodríguez who hasn't published it formally. As a result, information about it is hard to find. At the time of writing (early 2007), there is a discussion in file `new-entropy-coding-algorithm-312899.html` at web site `http://archives.devshed.com/forums/compression-130/` and some information (and source code) can also be obtained from this author.

The method employs phased-in codes, but is different from other entropy coders. It may perhaps be compared with static arithmetic coding. Most entropy coders assign variable-length codes to data symbols such that the length of the code for a symbol is inversely proportional to the symbol's frequency of occurrence. The RF method, in contrast, starts by assigning several fixed-length (block) codes to each symbol according to its probability. The method then associates a phased-in code (that the developer terms "redundant information") with each block codeword. Encoding is done in reverse, from the end of the input stream. Each symbol is replaced by one of its block codes $B$ in such a way that the phased-in code associated with $B$ is identical to some bits at the start (the leftmost part) of the compressed stream. Those bits are deleted from the compressed stream (which generates compression) and $B$ is prepended to it. For example, if the current block code is 010111 and the compressed stream is 0111|0001010..., then the result of prepending the code and removing identical bits is 01|0001010....

We start with an illustrative example. Given the four symbols $A$, $B$, $C$, and $D$, with probabilities 37.5%, 25%, 12.5%, and 25%, respectively, we assign each symbol several block codes according to its probability. The total number of codes must be a power of 2, so $A$ is assigned three codes, each of $B$ and $D$ gets two codes, and $C$ becomes the "owner" of one code, for a total of eight codes. Naturally, the codes are the 3-bit numbers 0 through 7. Table 1.11a lists the eight codes and their redundant information (the associated phased-in codes). Thus, e.g., the three codes of $A$ are associated with the phased-in codes 0, 10, and 11, because these are the codes for $n = 3$. (Section 1.9 shows that we have to look for the integer $m$ that satisfies $2^m \leq n < 2^{m+1}$. Thus, for $n = 3$, $m$ is 1. The first $2^m = 2$ symbols are assigned the 2-bit numbers $0 + 2$ and $1 + 2$ and the remaining $3 - 2$ symbol is assigned the 1-bit number $i - 2^m = 2 - 2 = 0$.) Similarly, the two phased-in codes associated with $B$ are 0 and 1. Symbol $D$ is associated with the same two codes, and the single block code 5 of $C$ has no associated phased-in codes because there are no phased-in codes for a set of one symbol. Table 1.11b is constructed similarly and lists 16 4-bit block codes and their associated phased-in codes for the three symbols $A$, $B$, and $C$ with probabilities 0.5, 0.2, and 0.2, respectively.

First, a few words on how to determine the number of codes per symbol from the number $n$ of symbols and their frequencies $f_i$. Given an input string of $F$ symbols (from an alphabet of $n$ symbols) such that symbol $i$ appears $f_i$ times (so that $\sum f_i = F$), we first determine the number of codes. This is simply the power $m$ of 2 that satisfies $2^{m-1} < n \leq 2^m$. We now multiply each $f_i$ by $2^m/F$. The new sum satisfies $\sum f_i \times 2^m/F = 2^m$. Next, we round each term of this sum to the nearest integer, and if any is rounded down to zero, we set it to 1. Finally, if the sum of these integers is slightly different from $2^m$, we increment (or decrement) each of the largest ones by 1 until the sum equals $2^m$.

| Code | Symbol | Redundant Info | Code | Symbol | Redundant Info |
|---|---|---|---|---|---|
| 0 | $A$ | $0/3 \to 0$ | 0 0000 | $A$ | $0/10 \to 000$ |
| 1 | $A$ | $1/3 \to 10$ | 1 0001 | $A$ | $1/10 \to 001$ |
| 2 | $A$ | $2/3 \to 11$ | 2 0010 | $A$ | $2/10 \to 010$ |
| 3 | $B$ | $0/2 \to 0$ | 3 0011 | $A$ | $3/10 \to 011$ |
| 4 | $B$ | $1/2 \to 1$ | 4 0100 | $A$ | $4/10 \to 100$ |
| 5 | $C$ | $0/1 \to -$ | 5 0101 | $A$ | $5/10 \to 101$ |
| 6 | $D$ | $0/2 \to 0$ | 6 0110 | $A$ | $6/10 \to 1100$ |
| 7 | $D$ | $1/2 \to 1$ | 7 0111 | $A$ | $7/10 \to 1101$ |
| | | | 8 1000 | $A$ | $8/10 \to 1110$ |
| | | | 9 1001 | $A$ | $8/10 \to 1111$ |
| | | | 10 1010 | $B$ | $0/3 \to 0$ |
| | | | 11 1011 | $B$ | $1/3 \to 10$ |
| | | | 12 1100 | $B$ | $2/3 \to 11$ |
| | | | 13 1101 | $C$ | $0/3 \to 0$ |
| | | | 14 1110 | $C$ | $1/3 \to 10$ |
| | | | 15 1111 | $C$ | $2/3 \to 11$ |

(a)                     (b)

Table 1.11: Eight and 16 RF Codes.

As an example, consider a 6-symbol alphabet and an input string of $F = 47$ symbols, where the six symbols appear 17, 6, 3, 12, 1, and 8 times. We first determine that $2^2 < 6 \le 2^3$, so we need 8 codes. Multiplying $17 \times 8/47 = 2.89 \to 3$, $6 \times 8/47 = 1.02 \to 1$, $3 \times 8/47 = 0.51 \to 1$, $12 \times 8/47 = 2.04 \to 2$, $1 \times 8/47 = 0.17 \to 0$, and $8 \times 8/47 = 1.36 \to 1$. The last step is to increase the 0 to 1, and make sure the sum is 8 by decrementing the largest count, 3, to 2.

The codes of Table 1.11b are now used to illustrate RF encoding. Assume that the input is the string $AABCA$. It is encoded from end to start. The last symbol $A$ is easy to encode as we can use any of its block codes. We therefore select 0000. The next symbol, $C$, has three block codes, and we select $13 = 1101$. The associated phased-in code is 0, so we start with 0000, delete the leftmost 0, and prepend 1101, to end up with $1101|000$. The next symbol is $B$ and we select block code $12 = 1100$ with associated phased-in code 11. Encoding is done by deleting the leftmost 11 and prepending 1100, to end up with $1100|01|000$. To encode the next $A$, we select block code $6 = 0110$ with associated phased-in code 1100. Again, we delete 1100 and prepend 0110 to end up with $0110||01|000$. Finally, the last (i.e., leftmost) symbol $A$ is reached, for which we select block code $3 = 0011$ (with associated phased-in code 011) and encode by deleting 011 and prepending 0011. The compressed stream is $0011|0||01|000$.

The RF encoding principle is simple. Out of all the block codes assigned to the current symbol, we select the one whose associated phased-in code is identical to the prefix of the compressed stream. This results in the deletion of the greatest number of bits and thus in maximum compression.

Decoding is the opposite of encoding. The decoder has access to the table of block codes and their associated codes (or it starts from the symbols' probabilities and constructs the table as the encoder does). The compressed stream is 0011001000 and the first code is the leftmost four bits $0011 = 3 \to A$. The first decoded symbol is $A$ and the decoder deletes the 0011 and prepends 011 (the phased-in code associated with 3) to end up with 011001000. The rest of the decoding is straightforward.

Experiments with this method verify that its performance is generally almost as good as Huffman coding. The main advantages of RF coding are as follows:

1. It works well with a 2-symbol alphabet. We know that Huffman coding fails in this situation, even when the probabilities of the symbols are skewed, because it simply assigns the two 1-bit codes to the symbols. In contrast, RF coding assigns the common symbol many (perhaps seven or 15) codes, while the rare symbol is assigned only one or two codes, thereby producing compression even in such a case.

2. The version of RF presented earlier is static. The probabilities of the symbols have to be known in advance in order for this version to work properly. It is possible to extend this version to a simple dynamic RF coding, where a buffer holds the most-recent symbols and code assignment is constantly modified. This version is described below.

3. It is possible to replace the phased-in codes with a simple form of arithmetic coding. This slows down both encoding and decoding, but results in better compression.

Dynamic RF coding is slower than the static version above, but is more efficient. Assuming that the probabilities of the data symbols are unknown in advance, this version of the basic RF scheme is based on a long sliding buffer. The buffer should be long, perhaps $2^{15}$ symbols or longer, in order to reflect the true frequencies of the symbols. A common symbol will tend to appear many times in the buffer and will therefore be assigned many codes. For example, given the alphabet A, B, C, D, and E, with probabilities 60%, 10%, 10%, 15%, and 5%, respectively, the buffer may, at a certain point in the encoding, hold the following

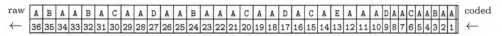

On the left, there is raw (unencoded) text and on the right there is text that has already been encoded. We can imagine the text being stationary and the buffer sliding to the left. If the buffer is long enough, the text inside it will reflect the true probabilities of the symbols and each symbol will have a number of codes proportional to its probability. At any time, the symbol immediately to the right of the buffer is encoded by selecting one of its codes in the buffer, and then moving the buffer one symbol to the left. If the buffer happens to contain no occurrences of the symbol to be encoded, then the code of all zeros is selected (which is why the codes in the buffer start at 1) and is output, followed by the raw (normally ASCII) code of the symbol. Notice that sliding the buffer modifies the codes of the symbols, but the decoder can do this in lockstep with the encoder. Once a code has been selected for a symbol, the code is prepended to the compressed stream after its associated phased-in code is used to delete identical bits, as in the static version.

We illustrate this version with a 4-symbol alphabet and the string ABBACBBBAADA. We assume a buffer with seven positions (so the codes are between 1 and 7) and place the buffer initially such that the rightmost A is immediately to its right, thus ABBA[CBBBAAD]A.

The initial buffer position and codes (both the 3-bit RF codes and the associated phased-in codes) are shown here. Symbol A is immediately to the right of the buffer and it can be encoded as either 2 or 3. We arbitrarily select 2, ending up with a compressed stream of 010.

| A B B A | C | B | B | B | A | A | D | A |
|---------|-----|-----|-----|-----|-----|-----|-----|---|
| | 7 | 6 | 5 | 4 | 3 | 2 | 1 | |
| | 111 | 110 | 101 | 100 | 011 | 010 | 001 | |
| | 1:- | 3:2 | 3:1 | 3:0 | 2:1 | 2:0 | 1:- | |
| | - | 11 | 10 | 0 | 1 | 0 | - | |

The buffer is slid, as shown below, thereby changing all the codes. This is why the dynamic version is slow. Symbol D is now outside the buffer and must be encoded as the pair (000, D) because there are no occurrences of D inside the buffer. The compressed stream becomes 000: 01000100|010.

| A B B | A | C | B | B | B | A | A | D A |
|-------|-----|-----|-----|-----|-----|-----|-----|-----|
| | 7 | 6 | 5 | 4 | 3 | 2 | 1 | |
| | 111 | 110 | 101 | 100 | 011 | 010 | 001 | |
| | 3:2 | 1:- | 3:2 | 3:1 | 3:0 | 3:1 | 3:0 | |
| | 11 | - | 11 | 10 | 0 | 10 | 0 | |

Now comes another A that can be encoded as either 1 or 6. Selecting the 1 also selects its associated phased-in code 0, so the leftmost 0 is deleted from the compressed stream and 001 is prepended. The result is 001|00: 01000100|010.

| A B | B | A | C | B | B | B | A | A D A |
|-----|-----|-----|-----|-----|-----|-----|-----|-----|
| | 7 | 6 | 5 | 4 | 3 | 2 | 1 | |
| | 111 | 110 | 101 | 100 | 011 | 010 | 001 | |
| | 4:3 | 2:1 | 1:- | 4:2 | 4:1 | 4:0 | 2:0 | |
| | 11 | 1 | - | 10 | 01 | 00 | 0 | |

The next symbol to be encoded is the third A from the right. The only available code is 5, which has no associated phased-in code. The output therefore becomes 101|001|00: 01000100|010.

| A | B | B | A | C | B | B | B | A A D A |
|---|-----|-----|-----|-----|-----|-----|-----|-----|
| | 7 | 6 | 5 | 4 | 3 | 2 | 1 | |
| | 111 | 110 | 101 | 100 | 011 | 010 | 001 | |
| | 5:4 | 5:3 | 1:- | 1:- | 5:2 | 5:1 | 5:0 | |
| | 111 | 110 | - | - | 10 | 01 | 00 | |

Next in line is the B. Four codes are available, of which the best choice is 5, with associated phased-in code 10. The string 101|1|001|00: 01000100|010 is the current output.

| A | B | B | A | C | B | B | B A A D A |
|---|-----|-----|-----|-----|-----|-----|-----|
| | 7 | 6 | 5 | 4 | 3 | 2 | 1 |
| | 111 | 110 | 101 | 100 | 011 | 010 | 001 |
| | 2:1 | 4:3 | 4:2 | 2:0 | 1:- | 4:1 | 4:0 |
| | 1 | 11 | 10 | 0 | - | 01 | 00 |

Encoding continues in this way even though the buffer is now only partially full. The next B is encoded with only three Bs in the buffer, and with each symbol encoded, fewer symbols remain in the buffer. Each time a symbol $s$ is encoded that has no copies left in the buffer, it is encoded as a pair of node 000 followed by the ASCII code of $s$. As the buffer gradually empties, more and more pairs are prepended to the output, thereby degrading the compression ratio. The last symbol (which is encoded with an empty buffer) is always encoded as a pair.

Thus, the decoder starts with an empty buffer, and reads the first code (000) which is followed by the ASCII code of the first (leftmost) symbol. That symbol is shifted into the buffer, and decoding continues as the reverse of encoding.

## 1.11 Recursive Phased-In Codes

The recursive phased-in codes were introduced in [Acharya and JáJá 95] and [Acharya and JáJá 96] as an enhancement to the well-known LZW (Lempel Ziv Welch) compression algorithm [Salomon 06]. These codes are easily explained in terms of complete binary trees, although their practical construction may be simpler with the help of certain recursive formulas conceived by Steven Pigeon.

The discussion in Section 2.18 shows that any positive integer $N$ can be written uniquely as the sum of certain powers of 2. Thus, for example, 45 is the sum $2^5 + 2^3 + 2^2 + 2^0$. In general, we can write $N = \sum_{i=1}^{s} 2^{a_i}$, where $a_1 > a_2 > \cdots > a_s \geq 0$ and $s \geq 1$. For $N = 45$, for example, these values are $s = 4$ and $a_1 = 5$, $a_2 = 3$, $a_3 = 2$, and $a_4 = 0$. Once a value for $N$ has been selected and the values of all its powers $a_i$ determined, a set of $N$ variable-length recursive phased-in codes can be constructed from the tree shown in Figure 1.12. For each power $a_i$, this tree has a subtree that is a complete binary tree of height $a_i$. The individual subtrees are connected to the root by $2s - 2$ edges labeled 0 or 1 as shown in the figure.

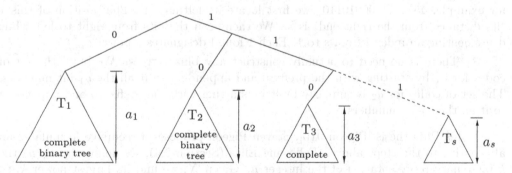

Figure 1.12: A Tree for Recursive Phased-In Codes.

The tree for $N = 45$ is shown in Figure 1.13. It is obvious that the complete binary tree for $a_i$ has $a_i$ leaves and the entire tree therefore has a total of $N$ leaf nodes. The codes are assigned by sliding down the tree from the root to each leaf, appending a 0 to the code each time we slide to the left and a 1 each time we slide to the right. Some codes are also shown in the figure. Thus, the 45 recursive phased-in codes for $N = 45$ are divided into the four sets $0xxxxx$, $10xxx$, $110xx$, and $111$, where the $x$'s stand for bits. The first set consists of the 32 5-bit combinations prepended by a 0, the second set includes eight 5-bit codes that start with 10, the third set has four codes, and the last set consists of the single code 111. As we scan the leaves of each subtree from left to right, we find that the codewords in each set are in ascending order. Even the codewords in different sets appear sorted, when scanned from left to right, if we append enough zeros to each so they all become six bits long. The codes are prefix codes, because, for example, once a code of the form $0xxxxx$ has been assigned, no other codes will start with 0.

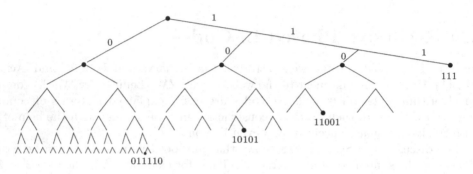

Figure 1.13: A Tree for $N = 45$.

In practice, these codes can be constructed in two steps, one trivial and the other one simple, as follows:

1. (This step is trivial.) Given a positive integer $N$, determine its powers $a_i$. Given, for example, $45 = \ldots 000101101$ we first locate its leftmost 1. The position of this 1 (its distance from the right end) is $s$. We then scan the bits from right to left while decrementing an index $i$ from $s$ to 1. Each 1 found designates a power $a_i$.

2. There is no need to actually construct any binary trees. We build the set of codes for $a_1$ by starting with the prefix 0 and appending to it all the $a_1$-bit numbers. The set of codes for $a_2$ is similarly built by starting with the prefix 10 and appending to it all the $a_2$-bit numbers.

In his PhD thesis [Pigeon 01b], Steven Pigeon proposes a recursive formula as an alternative to the steps above. Following Elias (Section 2.4), we denote by $\beta_k(n)$ the $k$-bit binary representation of the integer $n$. Given $N$, we find its largest power $k$, so $N = 2^k + b$ where $0 \le b < 2^k$ ($k$ equals $a_1$ above). The $N$ recursive phased-in codes $C_N(n)$ for $n = 0, 1, \ldots, N - 1$ are computed by

$$C_N(n) = \begin{cases} 0 : C_b(n), & \text{if } 0 \le n < b, \\ \beta_k(n), & \text{if } b = 0, \\ 1 : \beta_k(n - b), & \text{otherwise.} \end{cases}$$

Their lengths $L_N(n)$ are given by

$$L_N(n) = \begin{cases} 1 + L_b(n), & \text{if } 0 \le n < b, \\ k, & \text{if } b \le n, \\ 1 + k, & \text{otherwise.} \end{cases}$$

Table 1.14 lists the resulting codes for $N = 11$, 13, and 18. It is obvious that these are slightly different from the original codes of Acharya and JáJá. The latter code for $N = 11$, for example, consists of the sets $0xxx$, $10x$, and $11$, while Pigeon's formula generates the sets $1xxx$, $01x$, and $00$.

| $n$ | 11 | 13 | 18 |
|---|---|---|---|
| 0 | 00 | 00 | 00 |
| 1 | 010 | 0100 | 01 |
| 2 | 011 | 0101 | 10000 |
| 3 | 1000 | 0110 | 10001 |
| 4 | 1001 | 0111 | 10010 |
| 5 | 1010 | 1000 | 10011 |
| 6 | 1011 | 1001 | 10100 |
| 7 | 1100 | 1010 | 10101 |
| 8 | 1101 | 1011 | 10110 |
| 9 | 1110 | 1100 | 10111 |
| 10 | 1111 | 1101 | 11000 |
| 11 | | 1110 | 11001 |
| 12 | | 1111 | 11010 |
| 13 | | | 11011 |
| 14 | | | 11100 |
| 15 | | | 11101 |
| 16 | | | 11110 |
| 17 | | | 11111 |

Table 1.14: Codes for 11, 13, and 18.

The recursive phased-in codes bear a certain resemblance to the start-step-stop codes of Section 2.2, but a quick glance at Table 2.3 shows the difference between the two types of codes. A start-step-stop code consists of sets of codewords that start with 0, 10, 110, and so on and get longer and longer, while the recursive phased-in codes consist of sets that start with the same prefixes but get shorter.

> "The grand aim of all science is to cover the greatest number of empirical facts by logical deduction from the smallest number of hypotheses or axioms," he [Einstein] maintained. The same principle is at work in Ockham's razor, in Feynman's panegyric upon the atomic doctrine, and in the technique of data compression in information technology—all three of which extol economy of expression, albeit for different reasons.
>
> —Hans Christian von Baeyer, *Information, The New Language of Science* (2004)

# 1.12 Self-Delimiting Codes

Before we look at the main classes of VLCs, we list in this short section a few simple techniques (the first few of which are due to [Chaitin 66]) to construct *self-delimiting codes*, codes that have variable lengths and can be decoded unambiguously.

    1. Double each bit of the original message, so the message becomes a set of pairs of identical bits, then append a pair of different bits. Thus, the message 01001 becomes the bitstring 00|11|00|00|11|01. This is simple but is obviously too long. It is also fragile, because one bad bit will confuse any decoder (computer or human). A variation of this technique precedes each bit of the number with an intercalary bit of 1, except the last bit, which is preceded with a 0. Thus, 01001 become $1\bar{0}1\bar{1}1\bar{0}1\bar{0}0\bar{1}$. We can also concentrate the intercalary bits together and have them followed by the number, as in 11110|01001 (which is the number itself preceded by its unary code).

    2. Prepare a header with the length of the message and prepend it to the message. The size of the header depends on the size of the message, so the header should be made self-delimiting using method 1 above. Thus, the 6-bit message 010011 becomes the header 00|11|11|00|01 followed by 010011. It seems that the result is still very long (16 bits to encode six bits), but this is because our message is so short. Given a 1-million bit message, its length requires 20 bits. The self-delimiting header is therefore 42 bits long, increasing the length of the original message by 0.0042%.

    3. If the message is extremely long (trillions of bits) its header may become too long. In such a case, we can make the header itself self-delimiting by writing it in raw format and preceding it with its own header, which is made self-delimiting with method 1.

    4. It is now clear that there may be any number of headers. The first header is made self-delimiting with method 1, and all the other headers are concatenated to it in raw format. The last component is the (very long) original binary message.

    5. A decimal variable-length integer can be represented in base 15 (quindecimal) as a string of nibbles (groups of four bits each), where each nibble is a base-15 digit (i.e., between 0 and 14) and the last nibble contains $16 = 1111_2$. This method is sometimes referred to as nibble code or byte coding. Table 2.21 lists some examples.

    6. A variation on the nibble code is to start with the binary representation of the integer $n$ (or $n - 1$), prepend it with zeros until the total number of bits is divisible by 3, break it up into groups of three bits each, and prefix each group with a 0, except the leftmost (or alternatively, the rightmost) group, which is prepended by a 1. The length of this code for the integer $n$ is $4\lceil(\log_2 n)/3\rceil$, so it is ideal for a distribution of the form

$$2^{-4\lceil(\log_2 n)/3\rceil} \approx \frac{1}{n^{4/3}}. \tag{1.5}$$

This is a power law distribution with a parameter of 3/4. A natural extension of this code is to $k$-bit groups. Such a code fits power law distributions of the form

$$\frac{1}{n^{1+\frac{1}{k}}}. \tag{1.6}$$

    7. If the data to be compressed consists of a large number of small positive integers, then a word-aligned packing scheme may provide good (although not the best) compression combined with fast decoding. The idea is to pack several integers into fixed-size

fields of a computer word. Thus, if the word size is 32 bits, 28 bits may be partitioned into several $k$-bit fields while the remaining four bits become a selector that indicates the value of $k$.

The method described here is due to [Anh and Moffat 05] who employ it to compress inverted indexes. The integers they compress are positive and small because they are differences of consecutive pointers that are in sorted order. The authors describe three packing schemes, of which only the first, dubbed simple-9, is discussed here.

Simple-9 packs several small integers into 28 bits of a 32-bit word, leaving the remaining four bits as a selector. If the next 28 integers to be compressed all have values 1 or 2, then each can fit in one bit, making it possible to pack 28 integers in 28 bits. If the next 14 integers all have values of 1, 2, 3, or 4, then each fits in a 2-bit field and 14 integers can be packed in 28 bits. At the other extreme, if the next integer happens to be greater than $2^{14} = 16,384$, then the entire 28 bits must be devoted to it, and the 32-bit word contains just this integer. The choice of 28 is particularly fortuitous, because 28 is divisible by 1, 2, 3, 4, 5, 7, 9, 14, and itself. Thus, a 32-bit word packed in simple-9 may be partitioned in nine ways. Table 1.15 lists these nine partitions and shows that at most three bits are wasted (in row **e**).

| Selector | Number of codes | Code length | Unused bits |
|---|---|---|---|
| a | 28 | 1 | 0 |
| b | 14 | 2 | 0 |
| c | 9 | 3 | 1 |
| d | 7 | 4 | 0 |
| e | 5 | 5 | 3 |
| f | 4 | 7 | 0 |
| g | 3 | 9 | 1 |
| h | 2 | 14 | 0 |
| i | 1 | 28 | 0 |

Table 1.15: Summary of the Simple-9 Code.

Given the 14 integers 4, 6, 1, 1, 3, 5, 1, 7, 1, 13, 20, 1, 12, and 20, we encode the first nine integers as **c**|011|101|000|000|010|100|000|110|000|$b$ and the following five integers as **e**|01100|10011|00000|01011|10011|$bbb$, for a total of 64 bits, where each $b$ indicates an unused bit. The originators of this method point out that the use of a Golomb code would have compressed the 14 integers into 58 bits, but the small loss of compression efficiency of simple-9 is often more than compensated for by the speed of decoding. Once the leftmost four bit of a 32-bit word are examined and the selector value is determined, the remaining 28 bits can be unpacked with a few simple operations.

Allocating four bits for the selector is somewhat wasteful, because only nine of the 16 possible values are used, but the flexibility of the simple-9 code is the result of the many (nine) factors of 28. It is possible to give up one selector value, cut the selector size to three bits and increase the data segment to 29 bits, but 29 is a prime number, so a 29-bit segment cannot be partitioned into equal-size fields. The authors propose dividing a 32-bit word into a 2-bit selector and a 30-bit segment for packing data. The integer

30 has 10 factors, so a table of the simple-10 code, similar to Table 1.15, would have 10 rows. The selector field, however, can specify only four different values, which is why the resulting code (not described here) is more complex and is denoted by relative-10 instead of simple-10.

> Intercalary: Inserted between other elements or parts; interpolated.

# 1.13 Huffman Coding

### David Huffman (1925–1999)

Being originally from Ohio, it is no wonder that Huffman went to Ohio State University for his BS (in electrical engineering). What is unusual was his age (18) when he earned it in 1944. After serving in the United States Navy, he went back to Ohio State for an MS degree (1949) and then to MIT, for a PhD (1953, electrical engineering).

That same year, Huffman joined the faculty at MIT. In 1967, he made his only career move when he went to the University of California, Santa Cruz as the founding faculty member of the Computer Science Department. During his long tenure at UCSC, Huffman played a major role in the development of the department (he served as chair from 1970 to 1973) and he is known for his motto "my products are my students." Even after his retirement, in 1994, he remained active in the department, teaching information theory and signal analysis courses.

Huffman made significant contributions in several areas, mostly information theory and coding, signal designs for radar and communications, and design procedures for asynchronous logical circuits. Of special interest is the well-known Huffman algorithm for constructing a set of optimal prefix codes for data with known frequencies of occurrence. At a certain point he became interested in the mathematical properties of "zero curvature" surfaces, and developed this interest into techniques for folding paper into unusual sculptured shapes (the so-called computational origami).

Huffman coding is a popular method for compressing data with variable-length codes. Given a set of data symbols and their frequencies of occurrence (or, equivalently, their probabilities), the method constructs a set of variable-length codewords with the shortest average length for the symbols. Huffman coding serves as the basis for several popular applications implemented on popular platforms. Some programs use just the Huffman method, while others use it as one step in a multistep compression process. The Huffman method [Huffman 52] is somewhat similar to the Shannon–Fano method, proposed independently by Claude Shannon and Robert Fano in the late 1940s ([Shannon 48] and [Fano 49]). It generally produces better codes, and like the Shannon–Fano method, it produces the best code when the probabilities of the symbols are negative powers of 2. The main difference between the two methods is that Shannon–Fano constructs its codes top to bottom (from the leftmost to the rightmost bits), while Huffman constructs a code tree from the bottom up (builds the codes from right to left).

Since its development in 1952 by D. Huffman, this method has been the subject of intensive research in data compression. The long discussion in [Gilbert and Moore 59] proves that the Huffman code is a minimum-length code in the sense that no other encoding has a shorter average length. An algebraic approach to constructing the Huffman code is introduced in [Karp 61]. In [Gallager 74], Robert Gallager shows that the redundancy of Huffman coding is at most $p_1 + 0.086$ where $p_1$ is the probability of the most-common symbol in the alphabet. The redundancy is the difference between the average Huffman codeword length and the entropy. Given a large alphabet, such as the set of letters, digits and punctuation marks used by a natural language, the largest symbol probability is typically around 15–20%, bringing the value of the quantity $p_1 + 0.086$ to around 0.1. This means that Huffman codes are at most 0.1 bit longer (per symbol) than an ideal entropy encoder, such as arithmetic coding.

The Huffman algorithm starts by building a list of all the alphabet symbols in descending order of their probabilities. It then constructs a binary tree, the Huffman code tree, with a symbol at every leaf, from the bottom up. This is done in steps, where at each step two symbols with the smallest probabilities are selected, added to the top of the partial tree, deleted from the list, and replaced with an auxiliary symbol representing the two original symbols. When the list is reduced to just one auxiliary symbol (representing the entire alphabet), the tree is complete. The tree is then traversed to determine the codes of the individual symbols.

This process is best illustrated by an example. Given five symbols with probabilities as shown in Figure 1.16a, they are paired in the following order:

1. Symbol $a_4$ is combined with $a_5$ and both are replaced by the combined symbol $a_{45}$, whose probability is 0.2.

2. Four symbols are left, $a_1$, with probability 0.4, and $a_2$, $a_3$, and $a_{45}$, with probabilities 0.2 each. We arbitrarily select $a_3$ and $a_{45}$, combine them, and replace them with the auxiliary symbol $a_{345}$, whose probability is 0.4.

3. The three symbols $a_1$, $a_2$, and $a_{345}$, are now left, with probabilities 0.4, 0.2, and 0.4, respectively. We arbitrarily select $a_2$ and $a_{345}$, combine them, and replace them with the auxiliary symbol $a_{2345}$, whose probability is 0.6.

4. Finally, we combine the two remaining symbols, $a_1$ and $a_{2345}$, and replace them with $a_{12345}$ with probability 1.

The tree is now complete. It is shown in Figure 1.16a "lying on its side" with its root on the right and its five leaves on the left. To assign the codes, we arbitrarily assign a bit of 1 to the top edge, and a bit of 0 to the bottom edge, of every pair of edges. This results in the codewords 0, 10, 111, 1101, and 1100. The assignments of bits to the edges is arbitrary.

The average size of this code is $0.4 \times 1 + 0.2 \times 2 + 0.2 \times 3 + 0.1 \times 4 + 0.1 \times 4 = 2.2$ bits/symbol, but even more importantly, the Huffman code is not unique. Some of the steps above were chosen arbitrarily, since there were more than two symbols with the smallest probabilities. Figure 1.16b shows how the same five symbols can be combined differently to obtain a different Huffman code (11, 01, 00, 101, and 100). The average size of this code is $0.4 \times 2 + 0.2 \times 2 + 0.2 \times 2 + 0.1 \times 3 + 0.1 \times 3 = 2.2$ bits/symbol, the same as the previous code.

**Example.** Given the eight symbols A, B, C, D, E, F, G, and H with probabilities 1/30, 1/30, 1/30, 2/30, 3/30, 5/30, 5/30, and 12/30, we draw three different Huffman

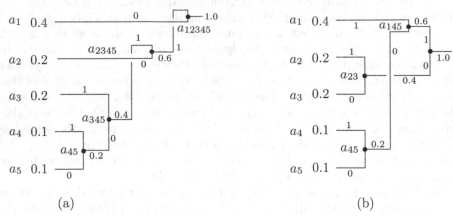

Figure 1.16: Two Equivalent Huffman Code Trees.

trees with heights 5 and 6 for these symbols and calculate the average code size for each tree. Figure 1.17a,b,c shows the three trees. The codes sizes for the trees are

$$(5 + 5 + 5 + 5\cdot2 + 3\cdot3 + 3\cdot5 + 3\cdot5 + 12)/30 = 76/30,$$
$$(5 + 5 + 4 + 4\cdot2 + 4\cdot3 + 3\cdot5 + 3\cdot5 + 12)/30 = 76/30,$$
$$(6 + 6 + 5 + 4\cdot2 + 3\cdot3 + 3\cdot5 + 3\cdot5 + 12)/30 = 76/30.$$

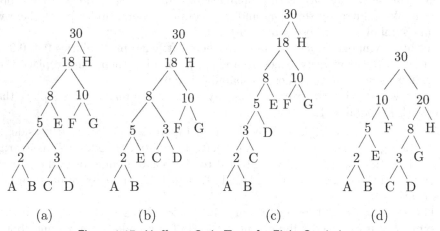

Figure 1.17: Huffman Code Trees for Eight Symbols.

As a self-exercise, consider the following question. Figure 1.17d shows another Huffman tree, with a height of 4, for the eight symbols introduced in the example above. Explain why this tree is wrong.

The answer is, after adding symbols A, B, C, D, E, F, and G to the tree, we were left with the three symbols ABEF (with probability 10/30), CDG (with probability

8/30), and H (with probability 12/30). The two symbols with lowest probabilities were ABEF and CDG, so they had to be merged. Instead, symbols CDG and H were merged, creating a non-Huffman tree.

The leaves of a Huffman code tree correspond to the individual codewords, but the interior nodes of the tree also play an important part. We already know that the codewords produced by the tree of Figure 1.16a are 0, 10, 111, 1101, and 1100. Once 0 has been assigned as a codeword, all other codewords must start with 1. Thus, 1 is a prefix of this code. Once 10 has been selected as a codeword, all the other codewords must start with 11. Thus, 11 is also a prefix of this code. Similarly, once 111 became a codeword, 110 became a prefix. Thus, the prefixes of this code are 1, 11, and 110, and it is easy to see that they correspond to nodes $a_{2345}$, $a_{345}$, and $a_{45}$ respectively. We can therefore say that the interior nodes of a Huffman code tree correspond to the prefixes of the code. It is often useful to claim that the root of the tree (node $a_{12345}$ in our case) corresponds to the empty prefix, which is sometimes denoted by $\Lambda$. The fast Huffman decoder of Section 1.13.3 is based on the code prefixes.

It turns out that the arbitrary decisions made in constructing the Huffman tree affect the individual codes but not the average size of the code. Still, we have to answer the obvious question, which of the different Huffman codes for a given set of symbols is best? The answer, while not obvious, is simple: the best code is the one with the smallest variance. The variance of a code measures how much the sizes of the individual codes deviate from the average size. The variance of code 1.16a is

$$0.4(1 - 2.2)^2 + 0.2(2 - 2.2)^2 + 0.2(3 - 2.2)^2 + 0.1(4 - 2.2)^2 + 0.1(4 - 2.2)^2 = 1.36,$$

while the variance of code 1.16b is

$$0.4(2 - 2.2)^2 + 0.2(2 - 2.2)^2 + 0.2(2 - 2.2)^2 + 0.1(3 - 2.2)^2 + 0.1(3 - 2.2)^2 = 0.16.$$

Code 1.16b is therefore preferable (see below). A careful look at the two trees shows how to select the one we want. In the tree of Figure 1.16a, symbol $a_{45}$ is combined with $a_3$, whereas in the tree of 1.16b it is combined with $a_1$. The rule for constructing the code with the smallest variance is therefore: when there are more than two smallest-probability nodes, select the ones that are lowest and highest in the tree and combine them. This will combine symbols of low probability with ones of high probability, thereby reducing the total variance of the code.

If the encoder simply writes the compressed stream on a file, the variance of the code makes no difference. A small-variance Huffman code is preferable only in cases where the encoder *transmits* the compressed stream, as it is being generated, over a communications line. In such a case, a code with large variance causes the encoder to generate bits at a rate that varies all the time. Since the bits have to be transmitted at a constant rate, the encoder has to use a buffer. Bits of the compressed stream are entered into the buffer as they are being generated and are moved out of it at a constant rate, to be transmitted. It is easy to see intuitively that a Huffman code with zero variance will enter bits into the buffer at a constant rate, so only a short buffer will be needed. The larger the code variance, the more variable is the rate at which bits enter the buffer, requiring the encoder to use a larger buffer.

The following claim is sometimes found in the literature:

It can be shown that the size of the Huffman code of a symbol
$a_i$ with probability $P_i$ is always less than or equal to $\lceil -\log_2 P_i\rceil$.

Even though it is often correct, this claim is not true in general. It seems to be a wrong corollary drawn by some authors from the Kraft–McMillan inequality, Equation (1.3). I am indebted to Guy Blelloch for pointing this out and also for the example of Table 1.18. In this example, the size of the Huffman code of a symbol $a_i$ is greater than $\lceil -\log_2 P_i\rceil$. The symbol in the second row of the table (indicated by an asterisk) has a 3-bit Huffman code, but satisfies $\lceil -\log_2 0.3\rceil = \lceil 1.737\rceil = 2$.

| $P_i$ | Code | $-\log_2 P_i$ | $\lceil -\log_2 P_i\rceil$ |
|-------|------|---------------|----------------------------|
| .01   | 000  | 6.644         | 7                          |
| *.30  | 001  | 1.737         | 2                          |
| .34   | 01   | 1.556         | 2                          |
| .35   | 1    | 1.515         | 2                          |

Table 1.18: A Huffman Code Example.

Note. It seems that the size of a code must also depend on the number $n$ of symbols (the size of the alphabet). A small alphabet requires just a few codes, so they can all be short; a large alphabet requires many codes, so some must be long. This being so, how can we say that the size of the code of symbol $a_i$ depends just on its probability $P_i$?

The explanation is simple. Imagine a large alphabet where all the symbols have (about) the same probability. Since the alphabet is large, that probability will be small, resulting in long codes. Imagine the other extreme case, where certain symbols have high probabilities (and, therefore, short codes). Since the probabilities have to add up to 1, the remaining symbols will have low probabilities (and, therefore, long codes). We therefore see that the size of a code depends on the probability, but is indirectly affected by the size of the alphabet.

Figure 1.19 shows a Huffman code for the 26 letters of the English alphabet (see also Table 3.13).

As a self-exercise, the reader may calculate the average size, entropy, and variance of this code.

Example. We present the Huffman codes for equal probabilities. Figure 1.20 shows Huffman codes for 5, 6, 7, and 8 symbols with equal probabilities. In the case where $n$ is a power of 2, the codewords are simply the fixed-sized (block) codes of the symbols. In other cases, the codewords are very close to the block codes. This shows that symbols with equal probabilities do not benefit from variable-length codes. (This is another way of saying that random text cannot be compressed.) Table 1.21 shows the codes, their average sizes and variances.

This example shows that symbols with equal probabilities don't compress under the Huffman method. This is understandable, since strings of such symbols normally make random text, and random text does not compress. There may be special cases where strings of symbols with equal probabilities are not random and can be compressed. A good example is the string $a_1 a_1 \ldots a_1 a_2 a_2 \ldots a_2 a_3 a_3 \ldots$ in which each symbol appears in

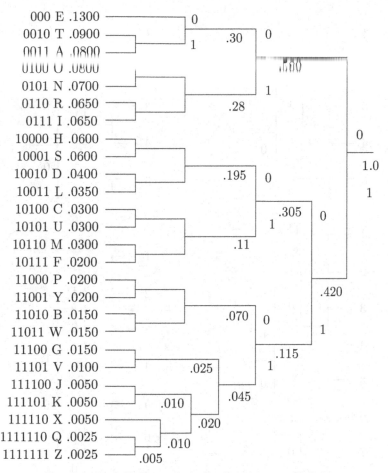

Figure 1.19: A Huffman Code for the 26-Letter Alphabet.

a long run. This string can be compressed with RLE (run-length encoding, Section 2.23) but not with Huffman codes.

Notice that the Huffman method cannot be applied to a two-symbol alphabet. In such an alphabet, one symbol is assigned the code 0 and the other is assigned code 1. The Huffman method cannot assign to any symbol a codeword shorter than one bit, so it cannot improve on this simple code. If the original data (the source) consists of individual bits, such as in the case of a bi-level (monochromatic) image, it is possible to combine several bits (perhaps four or eight) into a new symbol and pretend that the alphabet consists of these (16 or 256) symbols. The problem with this approach is that the original binary data may have certain statistical correlations between the bits, and some of these correlations would be lost when the bits are combined into symbols. When a typical bi-level image (a drawing or a diagram) is digitized by scan lines, a pixel is more likely to be followed by an identical pixel than by the opposite one. We therefore have a file that can start with either a 0 or a 1 (each has 0.5 probability

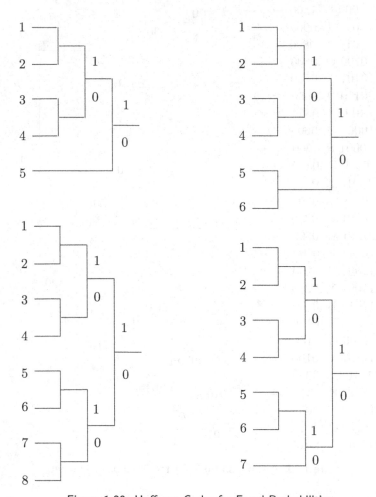

Figure 1.20: Huffman Codes for Equal Probabilities.

| $n$ | $p$ | $a_1$ | $a_2$ | $a_3$ | $a_4$ | $a_5$ | $a_6$ | $a_7$ | $a_8$ | Avg. size | Var. |
|---|---|---|---|---|---|---|---|---|---|---|---|
| 5 | 0.200 | 111 | 110 | 101 | 100 | 0 | | | | 2.6 | 0.64 |
| 6 | 0.167 | 111 | 110 | 101 | 100 | 01 | 00 | | | 2.672 | 0.2227 |
| 7 | 0.143 | 111 | 110 | 101 | 100 | 011 | 010 | 00 | | 2.86 | 0.1226 |
| 8 | 0.125 | 111 | 110 | 101 | 100 | 011 | 010 | 001 | 000 | 3 | 0 |

Table 1.21: Huffman Codes for 5–8 Symbols.

of being the first bit). A zero is more likely to be followed by another 0 and a 1 by another 1. Figure 1.22 is a finite-state machine illustrating this situation. If these bits are combined into, say, groups of eight, the bits inside a group will still be correlated, but the groups themselves will not be correlated by the original pixel probabilities. If the input stream contains, e.g., the two adjacent groups 00011100 and 00001110, they will be encoded independently, ignoring the correlation between the last 0 of the first group and the first 0 of the next group. Selecting larger groups improves this situation but increases the number of groups, which implies more storage for the code table and longer time to calculate the table.

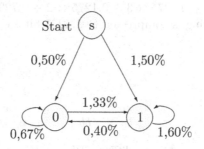

Figure 1.22: A Finite-State Machine.

Note. When the group size increases from $s$ bits to $s+n$ bits, the number of groups increases exponentially from $2^s$ to $2^{s+n} = 2^s \times 2^n$.

A more complex approach to image compression by Huffman coding is to create several complete sets of Huffman codes. If the group size is, e.g., eight bits, then several sets of 256 codes are generated. When a symbol S is to be encoded, one of the sets is selected, and S is encoded using its code in that set. The choice of set depends on the symbol preceding S.

**Example.** Given an image with 8-bit pixels where half the pixels have values 127 and the other half have values 128, we analyze the performance of RLE on the individual bitplanes of such an image, and compare it with what can be achieved with Huffman coding. The binary value of 127 is 01111111 and that of 128 is 10000000. Half the pixels in each bitplane will therefore be zeros and the other half will be 1's. In the worst case, each bitplane will be a checkerboard, i.e., will have many runs of size one. In such a case, each run requires a 1-bit code, leading to one codebit per pixel per bitplane, or eight codebits per pixel for the entire image, resulting in no compression at all. In comparison, a Huffman code for such an image requires just two codes (since there are just two pixel values) and they can be one bit each. This leads to one codebit per pixel, or a compression factor of eight.

## 1.13.1 Dual Tree Coding

Dual tree coding, an idea due to G. H. Freeman ([Freeman 91] and [Freeman 93]), combines Tunstall and Huffman coding in an attempt to improve the latter's performance for a 2-symbol alphabet. The idea is to use the Tunstall algorithm to extend such an alphabet from 2 symbols to $2^k$ strings of symbols, and select $k$ such that the probabilities of the strings will be close to negative powers of 2. Once this is achieved, the strings are assigned Huffman codes and the input stream is compressed by replacing the strings with these codes. This approach is illustrated here by a simple example.

Given a binary source that emits two symbols $a$ and $b$ with probabilities 0.15 and 0.85, respectively, we try to compress it in four different ways as follows:

1. We apply the Huffman algorithm directly to the two symbols. This simply assigns the two 1-bit codes 0 and 1 to $a$ and $b$, so there is no compression.

2. We combine the two symbols to obtain the four 2-symbol strings $aa$, $ab$, $ba$, and $bb$, with probabilities 0.0225, 0.1275, 0.1275, and 0.7225, respectively. The four strings are assigned Huffman codes as shown in Figure 1.23a, and it is obvious that the average code length is $0.0225 \times 3 + 0.1275 \times 3 + 0.1275 \times 2 + 0.7225 \times 1 = 1.4275$ bits. On average, each 2-symbol string is compressed to 1.4275 bits, yielding a compression ratio of $1.4275/2 \approx 0.714$.

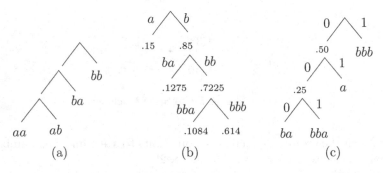

Figure 1.23: Dual Tree Coding.

3. We apply Tunstall's algorithm to obtain the four strings $bbb$, $bba$, $ba$, and $a$ with probabilities 0.614, 0.1084, 0.1275, and 0.15, respectively. The resulting parse tree is shown in Figure 1.23b. Tunstall's method compresses these strings by replacing each with a 2-bit code. Given a string of 257 bits with these probabilities, we expect the strings $bbb$, $bba$, $ba$, and $a$ to occur 61, 11, 13, and 15 times, respectively, for a total of 100 strings. Thus, Tunstall's method compresses the 257 input bits to $2 \times 100 = 200$ bits, for a compression ratio of $200/257 \approx 0.778$.

4. We now change the probabilities of the four strings above to negative powers of 2, because these are the best values for the Huffman method. Strings $bbb$, $bba$, $ba$, and $a$ are thus assigned the probabilities 0.5, 0.125, 0.125, and 0.25, respectively. The resulting Huffman code tree is shown in Figure 1.23c and it is easy to see that the 61, 11, 13, and 15 occurrences of these strings will be compressed to a total of $61 \times 1 + 11 \times 3 + 13 \times 3 + 15 \times 2 = 163$ bits, resulting in a compression ratio of $163/257 \approx 0.634$, much better.

To summarize, applying the Huffman method to a 2-symbol alphabet produces no compression. Combining the individual symbols in strings as in 2 above or applying the Tunstall method as in 3, produce moderate compression. In contrast, combining the strings produced by Tunstall with the codes generated by the Huffman method, results in much better performance. The dual tree method starts by constructing the Tunstall parse tree and then using its leaf nodes to construct a Huffman code tree. The only (still unsolved) problem is determining the best value of $k$. In our example, we iterated the Tunstall algorithm until we had $2^2 = 4$ strings, but iterating more times may have resulted in strings whose probabilities are closer to negative powers of 2.

## 1.13.2 Huffman Decoding

Before starting the compression of a data stream, the compressor (encoder) has to determine the codes. It does that based on the probabilities (or frequencies of occurrence) of the symbols. The probabilities or frequencies have to be written, as side information, on the compressed stream, so that any Huffman decompressor (decoder) will be able to decompress the stream. This is easy, since the frequencies are integers and the probabilities can be written as scaled integers. It normally adds just a few hundred bytes to the compressed stream. It is also possible to write the variable-length codes themselves on the stream, but this may be awkward, because the codes have different sizes. It is also possible to write the Huffman tree on the stream, but this may require more space than just the frequencies.

In any case, the decoder must know what information is supposed to be at the start of the stream, read it, and construct the Huffman tree for the alphabet. Only then can it read and decode the rest of the stream. The algorithm for decoding is simple. Start at the root and read the first bit off the compressed stream. If it is a 0, follow the bottom edge of the tree; if it is a 1, follow the top edge. Read the next bit and move another edge toward the leaves of the tree. When the decoder arrives at a leaf node, it finds the original, uncompressed code of the symbol (normally its ASCII code), and that code is emitted by the decoder. The process starts again at the root with the next bit.

This process is illustrated for the five-symbol alphabet of Figure 1.24. The four-symbol input string $a_4a_2a_5a_1$ is encoded into 1001100111. The decoder starts at the root, reads the first bit 1, and goes up. The second bit 0 sends it down, as does the third bit. This brings the decoder to leaf $a_4$, which it emits. It again returns to the root, reads 110, moves up, up, and down, to reach leaf $a_2$, and so on.

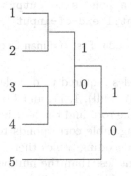

Figure 1.24: Huffman Decoding Illustrated.

## 1.13.3 Fast Huffman Decoding

Decoding a Huffman-compressed file by sliding down the code tree for each symbol is conceptually simple, but slow. The compressed file has to be read bit by bit and the decoder has to advance a node in the code tree for each bit. The method of this section, originally conceived by [Choueka et al. 85] but later reinvented by others, uses preset partial-decoding tables. These tables depend on the particular Huffman code used, but not on the data to be decoded. The compressed file is read in chunks of $k$ bits each (where $k$ is normally 8 or 16 but can have other values) and the current chunk is used as a pointer to a table. The table entry that is selected in this way can decode several symbols and it also points the decoder to the table to be used for the next chunk.

As an example, consider the Huffman code of Figure 1.16a, where the five codewords are 0, 10, 111, 1101, and 1100. The string of symbols $a_1a_1a_2a_4a_3a_1a_5\ldots$ is compressed by this code to the string $0|0|10|1101|111|0|1100\ldots$. We select $k=3$ and read this string in 3-bit chunks $001|011|011|110|110|0\ldots$. Examining the first chunk, it is easy to see that it should be decoded into $a_1a_1$ followed by the single bit 1 which is the prefix of another codeword. The first chunk is $001 = 1_{10}$, so we set entry 1 of the first table (table 0) to the pair $(a_1a_1, 1)$. When chunk 001 is used as a pointer to table 0, it points to entry 1, which immediately provides the decoder with the two decoded symbols $a_1a_1$ and also directs it to use table 1 for the next chunk. Table 1 is used when a partially-decoded chunk ends with the single-bit prefix 1. The next chunk is $011 = 3_{10}$, so entry 3 of table 1 corresponds to the encoded bits $1|011$. Again, it is easy to see that these should be decoded to $a_2$ and there is the prefix 11 left over. Thus, entry 3 of table 1 should be $(a_2, 2)$. It provides the decoder with the single symbol $a_2$ and also directs it to use table 2 next (the table that corresponds to prefix 11). The next chunk is again $011 = 3_{10}$, so entry 3 of table 2 corresponds to the encoded bits $11|011$. It is again obvious that these should be decoded to $a_4$ with a prefix of 1 left over. This process continues until the end of the encoded input. Figure 1.25 is the simple decoding algorithm in pseudocode.

```
i←0; output←null;
repeat
    j←input next chunk;
    (s,i)←Tableᵢ[j];
    append s to output;
until end-of-input
```

Figure 1.25: Fast Huffman Decoding.

Table 1.26 lists the four tables required to decode this code. It is easy to see that they correspond to the prefixes $\Lambda$ (null), 1, 11, and 110. A quick glance at Figure 1.16a shows that these correspond to the root and the four interior nodes of the Huffman code tree. Thus, each partial-decoding table corresponds to one of the four prefixes of this code. The number $m$ of partial-decoding tables therefore equals the number of interior nodes (plus the root) which is one less than the number $N$ of symbols of the alphabet.

Notice that some chunks (such as entry 110 of table 0) simply send the decoder to another table and do not provide any decoded symbols. Also, there is a tradeoff between chunk size (and thus table size) and decoding speed. Large chunks speed up decoding, but require large tables. A large alphabet (such as the 128 ASCII characters or the 256 8-bit bytes) also requires a large set of tables. The problem with large tables is that the decoder has to set up the tables after it has read the Huffman codes from the

| $T_0 = \Lambda$ | | | $T_1 = 1$ | | | $T_2 = 11$ | | | $T_3 = 110$ | | |
|---|---|---|---|---|---|---|---|---|---|---|---|
| 000 | $a_1a_1a_1$ | 0 | 1\|000 | $a_2a_1a_1$ | 0 | 11\|000 | $a_5a_1$ | 0 | 110\|000 | $a_5a_1a_1$ | 0 |
| 001 | $a_1a_1a_1$ | 1 | 1\|001 | $a_2a_1a_1$ | 1 | 11\|001 | $a_5$ | 1 | 110\|001 | $a_5a_1$ | 1 |
| 010 | $a_1a_2$ | 0 | 1\|010 | $a_2a_2$ | 0 | 11\|010 | $a_4a_1$ | 0 | 110\|010 | $a_5a_2$ | 0 |
| 011 | $a_1$ | 2 | 1\|011 | $a_2$ | 2 | 11\|011 | $a_4$ | 1 | 110\|011 | $a_5$ | 2 |
| 100 | $a_2a_1$ | 0 | 1\|100 | $a_5$ | 0 | 11\|100 | $a_3a_1a_1$ | 0 | 110\|100 | $a_4a_1a_1$ | 0 |
| 101 | $a_2$ | 1 | 1\|101 | $a_4$ | 0 | 11\|101 | $a_3a_1$ | 1 | 110\|101 | $a_4a_1$ | 1 |
| 110 | – | 3 | 1\|110 | $a_3a_1$ | 0 | 11\|110 | $a_3a_2$ | 0 | 110\|110 | $a_4a_2$ | 0 |
| 111 | $a_3$ | 0 | 1\|111 | $a_3$ | 1 | 11\|111 | $a_3$ | 2 | 110\|111 | $a_4$ | 2 |

Table 1.26: Partial-Decoding Tables for a Huffman Code.

compressed stream and before decoding can start, and this process may preempt any gains in decoding speed provided by the tables.

To set up the first table (table 0, which corresponds to the null prefix $\Lambda$), the decoder generates the $2^k$ bit patterns 0 through $2^k - 1$ (the first column of Table 1.26) and employs the decoding method of Section 1.13.2 to decode each pattern. This yields the second column of Table 1.26. Any remainders left are prefixes and are converted by the decoder to table numbers. They become the third column of the table. If no remainder is left, the third column is set to 0 (use table 0 for the next chunk). Each of the other partial-decoding tables is set in a similar way. Once the decoder decides that table 1 corresponds to prefix $p$, it generates the $2^k$ patterns $p|00\ldots0$ through $p|11\ldots1$ that become the first column of that table. It then decodes that column to generate the remaining two columns.

This method was conceived in 1985, when storage costs were considerably higher than today (early 2007). This prompted the developers of the method to find ways to cut down the number of partial-decoding tables, but these techniques are less important today and are not described here.

> Truth is stranger than fiction, but this is because fiction is obliged to stick to probability; truth is not.
>
> —Anonymous

## 1.13.4 Average Code Size

Figure 1.27a shows a set of five symbols with their probabilities and a typical Huffman code tree.

Symbol A appears 55% of the time and is assigned a 1-bit code, so it contributes $0.55 \cdot 1$ bits to the average code size. Symbol E appears only 2% of the time and is assigned a 4-bit Huffman code, so it contributes $0.02 \cdot 4 = 0.08$ bits to the code size. The average code size is therefore calculated to be

$$0.55 \cdot 1 + 0.25 \cdot 2 + 0.15 \cdot 3 + 0.03 \cdot 4 + 0.02 \cdot 4 = 1.7 \text{ bits per symbol.}$$

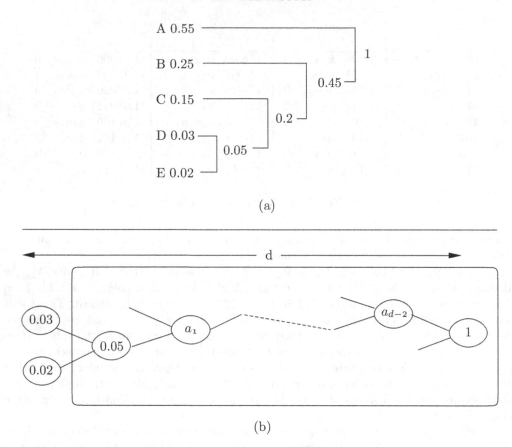

(a)

(b)

Figure 1.27: Huffman Code Trees.

Surprisingly, the same result is obtained by adding the values of the four internal nodes of the Huffman code tree $0.05 + 0.2 + 0.45 + 1 = 1.7$. This provides a way to compute the average code size of a set of Huffman codes without any multiplications. Simply add the values of all the internal nodes of the tree. Table 1.28 (where internal nodes are shown in italics) illustrates why this works. The left column consists of the values of all the internal nodes. The right columns show how each internal node is the sum of some of the leaf nodes. Summing the values in the left column yields 1.7, and summing the other columns shows that this 1.7 is the sum of the four values 0.02, the four values 0.03, the three values 0.15, the two values 0.25, and the single value 0.55.

This argument can be extended to the general case. It is easy to show that, in a Huffman-like tree (a tree where each node is the sum of its children), the weighted sum of the leaves, where the weights are the distances of the leaves from the root, equals the sum of the internal nodes. (This property has been communicated to me by John Motil.)

Figure 1.27b shows such a tree, where we assume that the two leaves 0.02 and 0.03 have $d$-bit Huffman codes. Inside the tree, these leaves become the children of internal node 0.05, which, in turn, is connected to the root by means of the $d - 2$ internal nodes

$$0.05 = \qquad\qquad = 0.02 + 0.03 + \cdots$$
$$a_1 \quad = 0.05 + \ldots = 0.02 + 0.03 + \cdots$$
$$a_2 \quad = a_1 \quad + \ldots = 0.02 + 0.03 + \cdots$$
$$\vdots \qquad -$$

$$.05 = \qquad\qquad .02 + .03$$
$$.20 = .05 + .15 = .02 + .03 + .15$$
$$.45 = .20 + .25 = .02 + .03 + .15 + .25$$
$$1.0 = .45 + .55 = .02 + .03 + .15 + .25 + .55$$

Table 1.28: Composition of Nodes.

$$a_{d-2} = a_{d-3} + \ldots = 0.02 + 0.03 + \cdots$$
$$1.0 \quad = a_{d-2} + \ldots = 0.02 + 0.03 + \cdots$$

Table 1.29: Composition of Nodes.

$a_1$ through $a_{d-2}$. Table 1.29 has $d$ rows and shows that the two values 0.02 and 0.03 are included in the various internal nodes exactly $d$ times. Adding the values of all the internal nodes produces a sum that includes the contributions $0.02 \cdot d + 0.03 \cdot d$ from the two leaves. Since these leaves are arbitrary, it is clear that this sum includes similar contributions from all the other leaves, so this sum is the average code size. Since this sum also equals the sum of the left column, which is the sum of the internal nodes, it is clear that the sum of the internal nodes equals the average code size.

Notice that this proof does not assume that the tree is binary. The property illustrated here exists for any tree where a node contains the sum of its children.

> "It needs compression," I suggested, cautiously.
> —Rudyard Kipling

## 1.13.5 Number of Codes

Since the Huffman code is not unique, a natural question is how many different codes are there? Figure 1.30a shows a Huffman code tree for six symbols, from which we can answer this question in two different ways.

Answer 1. The tree of Figure 1.30a has five interior nodes, and in general, a Huffman code tree for $n$ symbols has $n-1$ interior nodes. Each interior node has two edges coming out of it, labeled 0 and 1. Swapping the two labels produces a different Huffman code tree, so the total number of different Huffman code trees is $2^{n-1}$ (in our example, $2^5$ or 32). The tree of Figure 1.30b, for example, shows the result of swapping the labels of the two edges of the root. Table 1.31a,b lists the codes generated by the two trees.

Answer 2. The six codes of Table 1.31a can be divided into the four classes $00x$, $10y$, 01, and 11, where $x$ and $y$ are 1-bit each. It is possible to create different Huffman codes by changing the first two bits of each class. Since there are four classes, this is the same as creating all the permutations of four objects, something that can be done in $4! = 24$ ways. In each of the 24 permutations it is also possible to change the values of $x$ and $y$ in four different ways (since they are bits) so the total number of different Huffman codes in our six-symbol example is $24 \times 4 = 96$.

The two answers are different because they count different things. Answer 1 counts the number of different Huffman code trees, while answer 2 counts the number of different Huffman codes. It turns out that our example can generate 32 different code trees but only 94 different codes instead of 96. This shows that there are Huffman codes that cannot be generated by the Huffman method! Table 1.31c shows such an example. A look at the trees of Figure 1.30 should convince the reader that the codes of symbols 5

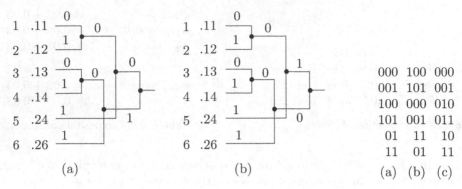

| | | |
|---|---|---|
| 000 | 100 | 000 |
| 001 | 101 | 001 |
| 100 | 000 | 010 |
| 101 | 001 | 011 |
| 01 | 11 | 10 |
| 11 | 01 | 11 |
| (a) | (b) | (c) |

Figure 1.30: Two Huffman Code Trees.          Table 1.31.

and 6 must start with different bits, but in the code of Table 1.31c they both start with 1. This code is therefore impossible to generate by any relabeling of the nodes of the trees of Figure 1.30.

## 1.13.6 Ternary Huffman Codes

The Huffman code is not unique. Moreover, it does not have to be binary! The Huffman method can easily be applied to codes based on other number systems (m-ary codes). Figure 1.32a shows a Huffman code tree for five symbols with probabilities 0.15, 0.15, 0.2, 0.25, and 0.25. The average code size is

$$2 \times 0.25 + 3 \times 0.15 + 3 \times 0.15 + 2 \times 0.20 + 2 \times 0.25 = 2.3 \, \text{bits/symbol}.$$

Figure 1.32b shows a ternary Huffman code tree for the same five symbols. The tree is constructed by selecting, at each step, three symbols with the smallest probabilities and merging them into one parent symbol, with the combined probability. The average code size of this tree is

$$2 \times 0.15 + 2 \times 0.15 + 2 \times 0.20 + 1 \times 0.25 + 1 \times 0.25 = 1.5 \, \text{trits/symbol}.$$

Notice that the ternary codes use the digits 0, 1, and 2.

**Example.** Given seven symbols with probabilities .02, .03, .04, .04, .12, .26, and .49, we construct binary and ternary Huffman code trees for them and calculate the average code size in each case. The two trees are shown in Figure 1.32c,d. The average code size for the binary Huffman tree is

$$1 \times 0.49 + 2 \times 0.25 + 5 \times 0.02 + 5 \times 0.03 + 5 \times .04 + 5 \times 0.04 + 3 \times 0.12 = 2 \, \text{bits/symbol},$$

and that of the ternary tree is

$$1 \times 0.26 + 3 \times 0.02 + 3 \times 0.03 + 3 \times 0.04 + 2 \times 0.04 + 2 \times 0.12 + 1 \times 0.49 = 1.34 \, \text{trits/symbol}.$$

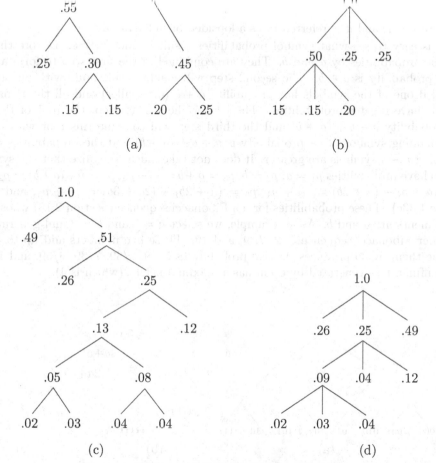

Figure 1.32: Binary and Ternary Huffman Code Trees.

## 1.13.7 Height of a Huffman Tree

The height of the code tree generated by the Huffman algorithm may sometimes be important because the height is also the length of the longest code in the tree. The popular Deflate method, for example, limits the lengths of certain Huffman codes to just 15 bits (because they have to fit in a 16-bit memory word or register).

It is easy to see that the shortest Huffman tree is created when the symbols have equal probabilities. If the symbols are denoted by A, B, C, and so on, then the algorithm combines pairs of symbols, such A and B, C and D, in the lowest level, and the rest of the tree consists of interior nodes as shown in Figure 1.33a. The tree is balanced or close to balanced and its height is $\lceil \log_2 n \rceil$. In the special case where the number of symbols $n$ is a power of 2, the height is exactly $\log_2 n$. In order to generate the tallest tree, we need to assign probabilities to the symbols such that each step in the Huffman method will increase the height of the tree by 1. Recall that each step in the Huffman algorithm combines two symbols. Thus, the tallest tree is obtained when the first step combines two of the $n$ symbols and each subsequent step combines the result of its predecessor with one of the remaining symbols (Figure 1.33b). The height of the final code tree is therefore $n - 1$, and it is referred to as a lopsided or unbalanced tree.

It is easy to see what symbol probabilities result in such a tree. Denote the two smallest probabilities by $a$ and $b$. They are combined in the first step to form a node whose probability is $a + b$. The second step will combine this node with an original symbol if one of the symbols has probability $a + b$ (or smaller) and all the remaining symbols have greater probabilities. Thus, after the second step, the root of the tree has probability $a + b + (a + b)$ and the third step will combine this root with one of the remaining symbols if its probability is $a + b + (a + b)$ and the probabilities of the remaining $n - 4$ symbols are greater. It does not take much to realize that the symbols have to have probabilities $p_1 = a$, $p_2 = b$, $p_3 = a + b = p_1 + p_2$, $p_4 = b + (a + b) = p_2 + p_3$, $p_5 = (a + b) + (a + 2b) = p_3 + p_4$, $p_6 = (a + 2b) + (2a + 3b) = p_4 + p_5$, and so on (Figure 1.33c). These probabilities form a Fibonacci sequence (Section 2.18) whose first two elements are $a$ and $b$. As an example, we select $a = 5$ and $b = 2$ and generate the 5-number Fibonacci sequence 5, 2, 7, 9, and 16. These five numbers add up to 39, so dividing them by 39 produces the five probabilities $5/39$, $2/39$, $7/39$, $9/39$, and $15/39$. The Huffman tree generated by them has a maximal height (which is 4).

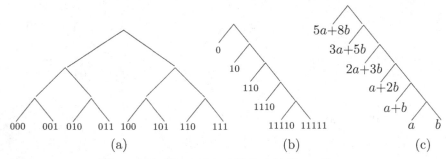

Figure 1.33: Shortest and Tallest Huffman Trees.

In principle, symbols in a set can have any probabilities, but in practice, the probabilities of symbols in an input file are computed by counting the number of occurrences of each symbol. Imagine a text file where only the nine symbols A through I appear. In order for such a file to produce the tallest Huffman tree, where the codes will have lengths from 1 to 8 bits, the frequencies of occurrence of the nine symbols have to form a Fibonacci sequence of probabilities. This happens when the frequencies of the symbols are 1, 1, 2, 3, 5, 8, 13, 21, and 34 (or integer multiples of these). The sum of these frequencies is 88, so our file has to be at least that long in order for a symbol to have 8-bit Huffman codes. Similarly, if we want to limit the sizes of the Huffman codes of a set of $n$ symbols to 16 bits, we need to count frequencies of at least 4180 symbols. To limit the code sizes to 32 bits, the minimum data size is 9,227,464 symbols.

If a set of symbols happens to have the Fibonacci probabilities and therefore results in a maximal-height Huffman tree with codes that are too long, the tree can be reshaped (and the maximum code length shortened) by slightly modifying the symbol probabil-ities so they are only much different from the original, but do not form a Fibonacci sequence.

## 1.13.8 Canonical Huffman Codes

The code of Table 1.31c has a simple interpretation. It assigns the first four symbols the 3-bit codes 0, 1, 2, 3, and the last two symbols the 2-bit codes 2 and 3. This is an example of a canonical Huffman code. Such a code has been selected from among the several (or even many) possible Huffman codes because its properties make it easy and fast to use and because it can be encoded more efficiently than the alternative codes.

| Canonical (adjective) |
| --- |
| 1. Of, relating to, or required by canon law. |
| 2. Of or appearing in the biblical canon. |
| 3. Conforming to orthodox or well-established rules or patterns, as of procedure. |
| 4. Of or belonging to a cathedral chapter. |
| 5. Of or relating to a literary canon. |
| 6. Music having the form of a canon. |

Table 1.34 shows a slightly bigger example of a canonical Huffman code. Imagine a set of 16 symbols (whose probabilities are irrelevant and are not shown) such that four symbols are assigned 3-bit codes, five symbols are assigned 5-bit codes, and the remaining seven symbols are assigned 6-bit codes. Table 1.34a shows a set of possible Huffman codes, while Table 1.34b shows a set of canonical Huffman codes. It is easy to see that the seven 6-bit canonical codes are simply the 6-bit integers 0 through 6. The five codes are the 5-bit integers 4 through 8, and the four codes are the 3-bit integers 3 through 6. We first show how these codes are generated and then how they are used.

| 1: | 000 | 011 | 9: | 10100 | 01000 |
| 2: | 001 | 100 | 10: | 101010 | 000000 |
| 3: | 010 | 101 | 11: | 101011 | 000001 |
| 4: | 011 | 110 | 12: | 101100 | 000010 |
| 5: | 10000 | 00100 | 13: | 101101 | 000011 |
| 6: | 10001 | 00101 | 14: | 101110 | 000100 |
| 7: | 10010 | 00110 | 15: | 101111 | 000101 |
| 8: | 10011 | 00111 | 16: | 110000 | 000110 |
| | (a) | (b) | | (a) | (b) |

Table 1.34.

| length: | 1 | 2 | 3 | 4 | 5 | 6 |
| --- | --- | --- | --- | --- | --- | --- |
| numl: | 0 | 0 | 4 | 0 | 5 | 7 |
| first: | 2 | 4 | 3 | 5 | 4 | 0 |

Table 1.35.

The top row of Table 1.35 (length) lists the possible code lengths, from 1 to 6 bits. The second row (numl) lists the number of codes of each length, and the bottom row (first) lists the first code in each group. This is why the three groups of codes start with values 3, 4, and 0. To obtain the top two rows we need to compute the lengths of all the Huffman codes for the given alphabet (see below). The third row is computed by setting "first[6]:=0;" and iterating

> for l:=5 downto 1 do first[l]:=⌈(first[l+1]+numl[l+1])/2⌉;

This guarantees that all the 3-bit prefixes of codes longer than three bits will be less than first[3] (which is 3), all the 5-bit prefixes of codes longer than five bits will be less than first[5] (which is 4), and so on. Once it is known how many codes are needed (and what the first code is) for each length, it is trivial to construct the full set of canonical codewords.

Now, for the applications of these unusual codes. Canonical Huffman codes are useful in cases where the alphabet is large and where fast decoding is mandatory. Because of the way the codes are constructed, it is easy for the decoder to identify the length of a code by reading and examining input bits one by one. Once the length is known, the symbol can be found in one step. The pseudocode listed here shows the rules for decoding:

> l:=1; input v;
> while v<first[l]
> append next input bit to v; l:=l+1;
> endwhile

As an example, suppose that the next code is 00110. As bits are input and appended to v, it goes through the values 0, 00=0, 001=1, 0011=3, 00110=6, while l is incremented from 1 to 5. All steps except the last satisfy v<first[l], so the last step determines the value of l (the code length) as 5. The symbol itself is found by subtracting v − first[5] = 6 − 4 = 2, so it is the third symbol (numbering starts from 0) in group l = 5 (symbol 7 of the 16 symbols).

It has been mentioned that canonical Huffman codes are useful in cases where the alphabet is large and fast decoding is important. A practical example is a collection of documents archived and compressed by a word-based adaptive Huffman coder. In an archive, a slow encoder is acceptable, but the decoder should be fast. When the individual symbols are words, the alphabet may be huge, making it impractical, or even impossible, to construct the Huffman code tree. However, even with a huge alphabet, the number of different code lengths is small, rarely exceeding 20 bits (just the number of 20-bit codes is about a million). If canonical Huffman codes are used, and the maximum code length is L, then the code length l of a symbol is found by the decoder in at most L steps, and the symbol itself is identified in one more step.

---

He uses statistics as a drunken man uses lampposts—for support rather than illumination.

              —Andrew Lang, *Treasury of Humorous Quotations*

---

The last point to be discussed is the encoder. In order to construct the canonical Huffman code, the encoder needs to know the length of the Huffman code of every symbol. The main problem is the large size of the alphabet, which may make it impractical or even impossible to build the entire Huffman code tree in memory. The algorithm presented here (see [Hirschberg and Lelewer 90] and [Sieminski 88]) solves this problem. It determines the code sizes for an alphabet of $n$ symbols using just one array of size

$2n$. One half of this array is used as a *heap*, so we start with a short description of this useful data structure.

A binary tree is a tree where every node has at most two children (i.e., it may have 0, 1, or 2 children). A complete binary tree is a binary tree where every node except the leaves has exactly two children. A balanced binary tree is a complete binary tree where some of the bottom-right nodes may be missing. A heap is a balanced binary tree where every leaf contains a data item and the items are ordered such that every path from a leaf to the root traverses nodes that are in sorted order, either nondecreasing (a max-heap) or nonincreasing (a min-heap). Figure 1.36 shows examples of min-heaps.

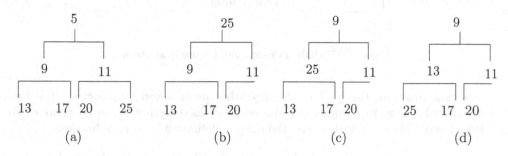

Figure 1.36: Min-Heaps.

A common operation on a heap is to remove the root and rearrange the remaining nodes to get back a heap. This is called *sifting* the heap. The four parts of Figure 1.36 show how a heap is sifted after the root (with data item 5) has been removed. Sifting starts by moving the bottom-right node to become the new root. This guarantees that the heap will remain a balanced binary tree. The root is then compared with its children and may have to be swapped with one of them in order to preserve the ordering of a heap. Several more swaps may be necessary to completely restore heap ordering. It is easy to see that the maximum number of swaps equals the height of the tree, which is $\lceil \log_2 n \rceil$.

The reason a heap must always remain balanced is that this makes it possible to store it in memory without using any pointers. The heap is said to be "housed" in an array. To house a heap in an array, the root is placed in the first array location (with index 1), the two children of the node at array location $i$ are placed at locations $2i$ and $2i + 1$, and the parent of the node at array location $j$ is placed at location $\lfloor j/2 \rfloor$. Thus the heap of Figure 1.36a is housed in an array by placing the nodes 5, 9, 11, 13, 17, 20, and 25 in the first seven locations of the array.

The algorithm uses a single array A of size $2n$. The frequencies of occurrence of the $n$ symbols are placed in the top half of A (locations $n + 1$ through $2n$), and the bottom half of A (locations 1 through $n$) becomes a min-heap whose data items are pointers to the frequencies in the top half (Figure 1.37a). The algorithm then goes into a loop where in each iteration the heap is used to identify the two smallest frequencies and replace them with their sum. The sum is stored in the last heap position A[h], and the heap shrinks by one position (Figure 1.37b). The loop repeats until the heap is reduced to just one pointer (Figure 1.37c).

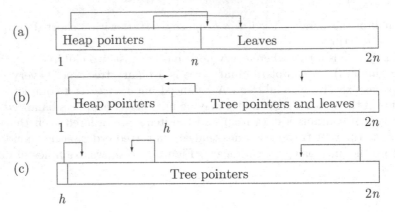

Figure 1.37: Huffman Heaps and Leaves in an Array.

We now illustrate this part of the algorithm using seven frequencies. The table below shows how the frequencies and the heap are initially housed in an array of size 14. Pointers are shown in italics, and the heap is delimited by square brackets.

| 1 | 2 | 3 | 4 | 5 | 6 | 7 | 8 | 9 | 10 | 11 | 12 | 13 | 14 |
|---|---|---|---|---|---|---|---|---|----|----|----|----|----|
| [14 | 12 | 13 | 10 | 11 | 9 | 8] | 25 | 20 | 13 | 17 | 9 | 11 | 5 |

The first iteration selects the smallest frequency (5), removes the root of the heap (pointer 14), and leaves A[7] empty.

| 1 | 2 | 3 | 4 | 5 | 6 | 7 | 8 | 9 | 10 | 11 | 12 | 13 | 14 |
|---|---|---|---|---|---|---|---|---|----|----|----|----|----|
| [12 | 10 | 13 | 8 | 11 | 9] | | 25 | 20 | 13 | 17 | 9 | 11 | 5 |

The heap is sifted, and its new root (12) points to the second smallest frequency (9) in A[12]. The sum $5 + 9$ is stored in the empty location 7, and the three array locations A[1], A[12], and A[14] are set to point to that location.

| 1 | 2 | 3 | 4 | 5 | 6 | 7 | 8 | 9 | 10 | 11 | 12 | 13 | 14 |
|---|---|---|---|---|---|---|---|---|----|----|----|----|----|
| [7 | 10 | 13 | 8 | 11 | 9] | 5+9 | 25 | 20 | 13 | 17 | 7 | 11 | 7 |

The heap is now sifted.

| 1 | 2 | 3 | 4 | 5 | 6 | 7 | 8 | 9 | 10 | 11 | 12 | 13 | 14 |
|---|---|---|---|---|---|---|---|---|----|----|----|----|----|
| [13 | 10 | 7 | 8 | 11 | 9] | 14 | 25 | 20 | 13 | 17 | 7 | 11 | 7 |

The new root is 13, implying that the smallest frequency (11) is stored at A[13]. The root is removed, and the heap shrinks to just five positions, leaving location 6 empty.

| 1 | 2 | 3 | 4 | 5 | 6 | 7 | 8 | 9 | 10 | 11 | 12 | 13 | 14 |
|---|---|---|---|---|---|---|---|---|----|----|----|----|----|
| [10 | 11 | 7 | 8 | 9] | | 14 | 25 | 20 | 13 | 17 | 7 | 11 | 7 |

The heap is now sifted. The new root is 10, showing that the second smallest frequency, 13, is stored at A[10]. The sum $11 + 13$ is stored at the empty location 6, and the three locations A[1], A[13], and A[10] are set to point to 6.

| 1 | 2 | 3 | 4 | 5 | | 6 | 7 | 8 | 9 | 10 | 11 | 12 | 13 | 14 |
|---|---|---|---|---|---|---|---|---|---|---|---|---|---|---|
| [6 | 11 | 7 | 8 | 9] | | 11+13 | 14 | 25 | 20 | 6 | 17 | 7 | 6 | 7 |

Figure 1.38 shows how the loop continues until the heap shrinks to just one node that is the single pointer 2. This indicates that the total frequency (which happens to be 100 in our example) is stored in A[2]. All other frequencies have been replaced by pointers. Figure 1.39a shows the heaps generated during the loop.

The final result of the loop is

| 1 | 2 | 3 | 4 | 5 | 6 | 7 | 8 | 9 | 10 | 11 | 12 | 13 | 14 |
|---|---|---|---|---|---|---|---|---|---|---|---|---|---|
| [2] | 100 | 2 | 2 | 3 | 4 | 5 | 3 | 4 | 6 | 5 | 7 | 6 | 7 |

from which it is easy to figure out the code lengths of all seven symbols. To find the length of the code of symbol 14, e.g., we follow the pointers 7, 5, 3, 2 from A[14] to the root. Four steps are necessary, so the code length is 4.

The code lengths for the seven symbols are 2, 2, 3, 3, 4, 3, and 4 bits. This can also be verified from the Huffman code tree of Figure 1.39b. A set of codes derived from this tree is shown in the following table:

| Count: | 25 | 20 | 13 | 17 | 9 | 11 | 5 |
|---|---|---|---|---|---|---|---|
| Code: | 01 | 11 | 101 | 000 | 0011 | 100 | 0010 |
| Length: | 2 | 2 | 3 | 3 | 4 | 3 | 4 |

## 1.13.9 Is Huffman Coding Dead?

The advantages of arithmetic coding are well known to users of compression algorithms. Arithmetic coding can compress data to its entropy, its adaptive version works well if fed the correct probabilities, and its performance does not depend on the size of the alphabet. On the other hand, arithmetic coding is slower than Huffman coding, its compression potential is not always utilized to its maximum, its adaptive version is very sensitive to the symbol probabilities and in extreme cases may even expand the data. Finally, arithmetic coding is not robust; a single error may propagate indefinitely and may result in wrong decoding of a large quantity of compressed data. (Some users may complain that they don't understand arithmetic coding and have no idea how to implement it, but this doesn't seem a serious concern, because implementations of this method are available for all major computing platforms.) A detailed comparison and analysis of both methods is presented in [Bookstein and Klein 93], with the conclusion that arithmetic coding has the upper hand only in rare situations.

In [Gallager 74], Robert Gallager shows that the redundancy of Huffman coding is at most $p_1 + 0.086$ where $p_1$ is the probability of the most-common symbol in the alphabet. The redundancy is the difference between the average Huffman codeword length and the entropy. Since arithmetic coding can compress data to its entropy, the quantity $p_1 + 0.086$ indicates by how much arithmetic coding outperforms Huffman coding. Given a 2-symbol alphabet, the more probable symbol appears with probability 0.5 or more, but given a large alphabet, such as the set of letters, digits and punctuation marks used by a language, the largest symbol probability is typically around 15–20%, bringing the value of the quantity $p_1 + 0.086$ to around 0.1. This means that Huffman codes are

```
 1   2   3   4   5   6   7   8   9  10  11  12  13  14
[7  11   6   8   9] 24  14  25  20   6  17   7   6   7

 1   2   3   4   5   6   7   8   9  10  11  12  13  14
[11  9   8   6]     24  14  25  20   6  17   7   6   7

 1   2   3   4       5   6   7   8   9  10  11  12  13  14
[11  9   8   6] 17+14  24  14  25  20   6  17   7   6   7

 1   2   3   4   5   6   7   8   9  10  11  12  13  14
[5   9   8   6] 31  24   5  25  20   6   5   7   6   7

 1   2   3   4   5   6   7   8   9  10  11  12  13  14
[9   6   8   5] 31  24   5  25  20   6   5   7   6   7

 1   2   3   4   5   6   7   8   9  10  11  12  13  14
[6   8   5]     31  24   5  25  20   6   5   7   6   7

 1   2   3       4   5   6   7   8   9  10  11  12  13  14
[6   8   5] 20+24  31  24   5  25  20   6   5   7   6   7

 1   2   3   4   5   6   7   8   9  10  11  12  13  14
[4   8   5] 44  31   4   5  25   4   6   5   7   6   7

 1   2   3   4   5   6   7   8   9  10  11  12  13  14
[8   5   4] 44  31   4   5  25   4   6   5   7   6   7

 1   2   3   4   5   6   7   8   9  10  11  12  13  14
[5   4]     44  31   4   5  25   4   6   5   7   6   7

 1   2       3   4   5   6   7   8   9  10  11  12  13  14
[5   4] 25+31  44  31   4   5  25   4   6   5   7   6   7

 1   2   3   4   5   6   7   8   9  10  11  12  13  14
[3   4] 56  44   3   4   5   3   4   6   5   7   6   7

 1   2   3   4   5   6   7   8   9  10  11  12  13  14
[4   3] 56  44   3   4   5   3   4   6   5   7   6   7

 1   2   3   4   5   6   7   8   9  10  11  12  13  14
[3]     56  44   3   4   5   3   4   6   5   7   6   7

 1       2   3   4   5   6   7   8   9  10  11  12  13  14
[3] 56+44  56  44   3   4   5   3   4   6   5   7   6   7

 1       2   3   4   5   6   7   8   9  10  11  12  13  14
[2] 100    2   2   3   4   5   3   4   6   5   7   6   7
```

Figure 1.38: Sifting the Heap.

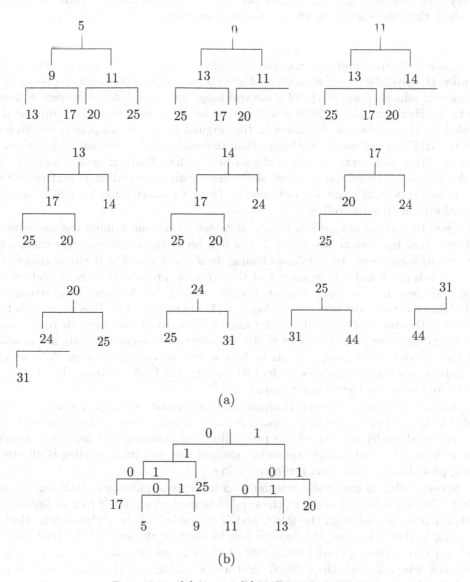

Figure 1.39: (a) Heaps. (b) Huffman Code Tree.

Considine's Law. Whenever one word or letter can change the entire meaning of a sentence, the probability of an error being made will be in direct proportion to the embarrassment it will cause.

—Bob Considine

at most 0.1 bit longer (per symbol) than arithmetic coding. For some (perhaps even many) applications, such a small difference may be insignificant, but those applications for which this difference is significant may be important.

Bookstein and Klein examine the two extreme cases of large and small alphabets. Given a text file in a certain language, it is often compressed in blocks. This limits the propagation of errors and also provides several entry points into the file. The authors examine the probabilities of characters of several large alphabets (each consisting of the letters and punctuations marks of a natural language), and list the average codeword length for Huffman and arithmetic coding (the latter is the size of the compressed file divided by the number of characters in the original file). The surprising conclusion is that the Huffman codewords are longer than the arithmetic codewords by less than one percent. Also, arithmetic coding performs better than Huffman coding only in large blocks of text. The minimum block size where arithmetic coding is preferable turns out to be between 269 and 457 characters. Thus, for shorter blocks, Huffman coding outperforms arithmetic coding.

The other extreme case is a binary alphabet where one symbol has probability $e$ and the other has probability $1 - e$. If $e = 0.5$, no method will compress the data. If the probabilities are skewed, Huffman coding does a bad job. The Huffman codes of the two symbols are 0 and 1 independent of the symbols' probabilities. Each code is 1-bit long, and there is no compression. Arithmetic coding, on the other hand, compresses such data to its entropy, which is $-[e \log_2 e + (1 - e) \log_2(1 - e)]$. This expression tends to 0 for both small $e$ (close to 0) and for large $e$ (close to 1). However, there is a simple way to improve the performance of Huffman coding in this case. Simply group several bits into a word. If we group the bits in 4-bit words, we end up with an alphabet of 16 symbols, where the probabilities are less skewed and the Huffman codes do a better job, especially because of the Gallager bound.

Another difference between Huffman and arithmetic coding is the case of wrong probabilities. This is especially important when a compression algorithm employs a mathematical model to estimate the probabilities of occurrence of individual symbols. The authors show that, under reasonable assumptions, arithmetic coding is affected by wrong probabilities more than Huffman coding.

Speed is also an important consideration in many applications. Huffman encoding is fast. Given a symbol to encode, the symbol is used as a pointer to a code table, the Huffman code is read from the table, and is appended to the codes-so-far. Huffman decoding is slower because the decoder has to start at the root of the Huffman code tree and slide down, guided by the bits of the current codeword, until it reaches a leaf node, where it finds the symbol. Arithmetic coding, on the other hand, requires multiplications and divisions, and is therefore slower. (Notice, however, that certain versions of arithmetic coding, most notably the Q-coder, MQ-coder, and QM-coder, have been developed specifically to avoid slow operations and are not slow.)

Often, a data compression application requires a certain amount of robustness against transmission errors. Neither Huffman nor arithmetic coding is robust, but it is known from long experience that Huffman codes tend to synchronize themselves fairly quickly following an error, in contrast to arithmetic coding, where an error may propagate to the end of the compressed file. It is also possible to construct resynchronizing Huffman codes, as shown in Section 3.4.

The conclusion is that Huffman coding, being fast, simple, and effective, is preferable to arithmetic coding for most applications. Arithmetic coding is the method of choice only in cases where the alphabet has skewed probabilities that cannot be redefined.

> In Japan, the basic codes are the Civil Code, the Commercial Code, the Penal Code, and procedural codes such as the Code of Criminal Procedure and the Code of Civil Procedure.
>
> —Roger E. Meiners, *The Legal Environment of Business*

# 2
# Advanced Codes

We start this chapter with codes for the integers. This is followed by many types of variable-length codes that are based on diverse principles, have been developed by different approaches, and have various useful properties and features.

## 2.1 VLCs for Integers

Following Elias, it is customary to denote the standard binary representation of the integer $n$ by $\beta(n)$. This representation can be considered a code (the beta code), but it does not satisfy the prefix property (because, for example, $2 = 10_2$ is the prefix of $4 = 100_2$). The beta code has another disadvantage. Given a set of integers between 0 and $n$, we can represent each in $1 + \lfloor \log_2 n \rfloor$ bits, a fixed-length representation. However, if the number of integers in the set is not known in advance (or if the largest integer is unknown), a fixed-length representation cannot be used and the natural solution is to assign variable-length codes to the integers. Any variable-length codes for integers should satisfy the following requirements:

1. Given an integer $n$, its code should be as short as possible and should be constructed from the magnitude, length, and bit pattern of $n$, without the need for any table lookups or other mappings.

2. Given a bitstream of variable-length codes, it should be easy to decode the next code and obtain an integer $n$ even if $n$ hasn't been seen before.

We will see that in many VLCs for integers, part of the binary representation of the integer is included in the code, and the rest of the code is side information indicating the length or precision of the encoded integer.

Several codes for the integers are described in the first few sections of this chapter. Some of them can code only nonnegative integers and others can code only positive integers. A VLC for positive integers can be extended to encode nonnegative integers by incrementing the integer before it is encoded and decrementing the result produced by decoding. A VLC for arbitrary integers can be obtained by a bijection, a mapping of the form

$$
\begin{array}{cccccccccccc}
0 & -1 & 1 & -2 & 2 & -3 & 3 & -4 & 4 & -5 & 5 & \cdots \\
1 & 2 & 3 & 4 & 5 & 6 & 7 & 8 & 9 & 10 & 11 & \cdots
\end{array}
$$

> A function is bijective if it is one-to-one and onto.

Perhaps the simplest variable-length code for integers is the well-known unary code. The unary code of the positive integer $n$ is constructed as $n - 1$ bits of 1 followed by a single 0, or alternatively as $n - 1$ zeros followed by a single 1 (the three left columns of Table 2.1). The length of the unary code for the integer $n$ is therefore $n$ bits. The two rightmost columns of Table 2.1 show how the unary code can be extended to encode the nonnegative integers (which makes the codes one bit longer). The unary code is simple to construct and is useful in many applications, but it is not universal. Stone-age people indicated the integer $n$ by marking $n$ adjacent vertical bars on a stone, which is why the unary code is sometimes known as a stone-age binary and each of its $n$ or $(n-1)$ 1's (or $n$ or $(n-1)$ zeros) is termed a stone-age bit.

Stone Age Binary?

| $n$ | Code | Reverse | Alt. code | Alt. reverse |
|---|---|---|---|---|
| 0 | – | – | 0 | 1 |
| 1 | 0 | 1 | 10 | 01 |
| 2 | 10 | 01 | 110 | 001 |
| 3 | 110 | 001 | 1110 | 0001 |
| 4 | 1110 | 0001 | 11110 | 00001 |
| 5 | 11110 | 00001 | 111110 | 000001 |

Table 2.1: Some Unary Codes.

It is easy to see that the unary code satisfies the prefix property, so it is instantaneous and can be used as a variable-length code. Since its length $L$ satisfies $L = n$, we get $2^{-L} = 2^{-n}$, so it makes sense to use this code in cases where the input data consists of integers $n$ with exponential probabilities $P(n) \approx 2^{-n}$. Given data that lends itself to the use of the unary code (i.e., a set of integers that satisfy $P(n) \approx 2^{-n}$), we can assign unary codes to the integers and these codes will be as good as the Huffman codes with the advantage that the unary codes are trivial to encode and decode. In general, the unary code is used as part of other, more sophisticated, variable-length codes.

**Example**: Table 2.2 lists the integers 1 through 6 with probabilities $P(n) = 2^{-n}$, except that $P(6) = 2^{-5} \approx 2^{-6}$. The table lists the unary codes and Huffman codes for the six integers, and it is obvious that these codes have the same lengths (except the code of 6, because this symbol does not have the correct probability).

> Every positive number was one of Ramanujan's personal friends.
>                                                    —J. E. Littlewood

| $n$ | Prob. | Unary | Huffman |
|---|---|---|---|
| 1 | $2^{-1}$ | 0 | 0 |
| 2 | $2^{-2}$ | 10 | 10 |
| 3 | $2^{-3}$ | 110 | 110 |
| 4 | $2^{-4}$ | 1110 | 1110 |
| 5 | $2^{-5}$ | 11110 | 11110 |
| 6 | $2^{-5}$ | 111110 | 11111 |

Table 2.2: Six Unary and Huffman Codes.

## 2.2 Start-Step-Stop Codes

The unary code is ideal for compressing data that consists of integers $n$ with probabilities $P(n) \approx 2^{-n}$. If the data to be compressed consists of integers with different probabilities, it may benefit from one of the general unary codes (also known as start-step-stop codes). Such a code, proposed by [Fiala and Greene 89], depends on a triplet (start, step, stop) of nonnegative integer parameters. A set of such codes is constructed subset by subset as follows:

1. Set $n = 0$.
2. Set $a = \text{start} + n \times \text{step}$.
3. Construct the subset of codes that start with $n$ leading 1's, followed by a single intercalary bit (separator) of 0, followed by a combination of $a$ bits. There are $2^a$ such codes.
4. Increment $n$ by step. If $n < \text{stop}$, go to step 2. If $n > \text{stop}$, issue an error and stop. If $n = \text{stop}$, repeat steps 2 and 3 but without the single intercalary 0 bit of step 3, and stop.

This construction makes it obvious that the three parameters have to be selected such that $\text{start} + n \times \text{step}$ will reach "stop" for some nonnegative $n$. The number of codes for a given triplet is normally finite and depends on the choice of parameters. It is given by

$$\frac{2^{\text{stop}+\text{step}} - 2^{\text{start}}}{2^{\text{step}} - 1}.$$

Notice that this expression increases exponentially with parameter "stop," so large sets of these codes can be generated even with small values of the three parameters. Notice that the case step $= 0$ results in a zero denominator and thus in an infinite set of codes.

Tables 2.3 and 2.4 show the 680 codes of (3,2,9) and the 2044 codes of (2,1,10). Table 2.5 lists the number of codes of each of the general unary codes $(2, 1, k)$ for $k = 2, 3, \ldots, 11$. This table was calculated by the *Mathematica* command `Table[2^(k+1)-4,{k,2,11}]`.

**Examples:**

1. The triplet $(n, 1, n)$ defines the standard (beta) $n$-bit binary codes, as can be verified by direct construction. The number of such codes is easily seen to be

$$\frac{2^{n+1} - 2^n}{2^1 - 1} = 2^n.$$

| $n$ | $a =$ $3 + n \cdot 2$ | $n$th codeword | Number of codewords | Range of integers |
|---|---|---|---|---|
| 0 | 3 | $0xxx$ | $2^3 = 8$ | 0–7 |
| 1 | 5 | $10xxxxx$ | $2^5 = 32$ | 8–39 |
| 2 | 7 | $110xxxxxxx$ | $2^7 = 128$ | 40–167 |
| 3 | 9 | $111xxxxxxxxx$ | $2^9 = 512$ | 168–679 |
| | | Total | 680 | |

Table 2.3: The General Unary Code (3,2,9).

| $n$ | $a =$ $2 + n \cdot 1$ | $n$th codeword | Number of codewords | Range of integers |
|---|---|---|---|---|
| 0 | 2 | $0xx$ | 4 | 0–3 |
| 1 | 3 | $10xxx$ | 8 | 4–11 |
| 2 | 4 | $110xxxx$ | 16 | 12–27 |
| 3 | 5 | $1110xxxxx$ | 32 | 28–59 |
| | | $\cdots$ | | $\cdots$ |
| 8 | 10 | $\underbrace{11...1}_{8}\underbrace{xx...x}_{10}$ | 1024 | 1020–2043 |
| | | Total | 2044 | |

Table 2.4: The General Unary Code (2,1,10).

| $k$ : | 2 | 3 | 4 | 5 | 6 | 7 | 8 | 9 | 10 | 11 |
|---|---|---|---|---|---|---|---|---|---|---|
| $(2,1,k)$: | 4 | 12 | 28 | 60 | 124 | 252 | 508 | 1020 | 2044 | 4092 |

Table 2.5: Number of General Unary Codes $(2,1,k)$ for $k = 2, 3, \ldots, 11$.

2. The triplet $(0, 0, \infty)$ defines the codes 0, 10, 110, 1110,... which are the unary codes but assigned to the integers 0, 1, 2,... instead of 1, 2, 3,... .

3. The triplet $(0, 1, \infty)$ generates a variance of the Elias gamma code.

4. The triplet $(k, k, \infty)$ generates another variance of the Elias gamma code.

5. The triplet $(k, 0, \infty)$ generates the Rice code with parameter $k$.

6. The triplet $(s, 1, \infty)$ generates the exponential Golomb codes of page 164.

7. The triplet $(1, 1, 30)$ produces $(2^{30} - 2^1)/(2^1 - 1) \approx$ a billion codes.

8. Table 2.6 shows the general unary code for (10,2,14). There are only three code lengths since "start" and "stop" are so close, but there are many codes because "start" is large.

| $n$ | $a =$ $10 + n \cdot 2$ | $n$th codeword | Number of codewords | Range of integers |
|---|---|---|---|---|
| 0 | 10 | $0\underbrace{x...x}_{10}$ | $2^{10} = 1K$ | 0–1023 |
| 1 | 12 | $10\underbrace{xx...x}_{12}$ | $2^{12} = 4K$ | 1024–5119 |
| 2 | 14 | $11\underbrace{xx...xx}_{14}$ | $2^{14} = 16K$ | 5120–21503 |
| | | Total | 21504 | |

Table 2.6: The General Unary Code (10,2,14).

# 2.3 Start/Stop Codes

The start-step-stop codes are flexible. By carefully adjusting the values of the three parameters it is possible to construct sets of codes of many different lengths. However, the lengths of these codes are restricted to the values $n + 1 + \text{start} + n \times \text{step}$ (except for the last subset where the codes have lengths $\text{stop} + \text{start} + \text{stop} \times \text{step}$). The start/stop codes of this section were conceived by Steven Pigeon and are described in [Pigeon 01a,b], where it is shown that they are universal. A set of these codes is fully specified by an array of nonnegative integer parameters $(m_0, m_1, \ldots, m_t)$ and is constructed in subsets, similar to the start-step-stop codes, in the following steps:

1. Set $i = 0$ and $a = m_0$.
2. Construct the subset of codes that start with $i$ leading 1's, followed by a single separator of 0, followed by a combination of $a$ bits. There are $2^a$ such codes.
3. Increment $i$ by 1 and set $a \leftarrow a + m_i$.
4. If $i < t$, go to step 2. Otherwise ($i = t$), repeat step 2 but without the single 0 intercalary, and stop.

Thus, the parameter array $(0, 3, 1, 2)$ results in the set of codes listed in Table 2.7.

| $i$ | $a$ | Codeword | # of codes | Length |
|---|---|---|---|---|
| 0 | 2 | $0xx$ | 4 | 3 |
| 1 | 5 | $10xxxxx$ | 32 | 7 |
| 2 | 6 | $110xxxxxx$ | 64 | 9 |
| 3 | 8 | $111xxxxxxxx$ | 256 | 11 |

Table 2.7: The Start/Stop Code (2,3,1,2).

The maximum code length is $t + m_0 + \cdots + m_t$ and on average the start/stop code of an integer $s$ is never longer than $\lceil \log_2 s \rceil$ (the length of the binary representation of $s$). If an optimal set of such codes is constructed by an encoder and is used to compress a data file, then the only side information needed in the compressed file is the value of $t$ and the array of $t + 1$ parameters $m_i$ (which are mostly small integers). This is considerably less than the size of a Huffman tree or the side information required by many other compression methods.

The start/stop codes can also encode an indefinite number of arbitrarily large integers. Simply set all the parameters to the same value and set $t$ to infinity.

Steven Pigeon, the developer of these codes, shows that the parameters of the start/stop codes can be selected such that the resulting set of codes will have an average length shorter than what can be achieved with the start-step-stop codes for the same probability distribution. He also shows how the probability distribution can be employed to determine the best set of parameters for the code. In addition, the number of codes in the set can be selected as needed, in contrast to the start-step-stop codes that often result in more codes than needed.

> The Human Brain starts working the moment you are born and never stops until you stand up to speak in public!
>
> —George Jessel

# 2.4 Elias Codes

In his pioneering work [Elias 75], Peter Elias described three useful prefix codes. The main idea of these codes is to prefix the integer being encoded with an encoded representation of its order of magnitude. For example, for any positive integer $n$ there is an integer $M$ such that $2^M \le n < 2^{M+1}$. We can therefore write $n = 2^M + L$ where $L$ is at most $M$ bits long, and generate a code that consists of $M$ and $L$. The problem is to determine the length of $M$ and this is solved in different ways by the various Elias codes. Elias denoted the unary code of $n$ by $\alpha(n)$ and the standard binary representation of $n$, from its most-significant 1, by $\beta(n)$. His first code was therefore designated $\gamma$ (gamma).

**Elias gamma code** $\gamma(n)$ for positive integers $n$ is simple to encode and decode and is also universal.

**Encoding.** Given a positive integer $n$, perform the following steps:

1. Denote by $M$ the length of the binary representation $\beta(n)$ of $n$.
2. Prepend $M - 1$ zeros to it (i.e., the $\alpha(n)$ code without its terminating 1).

Step 2 amounts to prepending the length of the code to the code, in order to ensure unique decodability.

The length $M$ of the integer $n$ is, from Equation (1.1), $1 + \lfloor \log_2 n \rfloor$, so the length of $\gamma(n)$ is

$$2M - 1 = 2\lfloor \log_2 n \rfloor + 1. \tag{2.1}$$

We later show that this code is ideal for applications where the probability of $n$ is $1/(2n^2)$.

An alternative construction of the gamma code is as follows:

1. Find the largest integer $N$ such that $2^N \le n < 2^{N+1}$ and write $n = 2^N + L$. Notice that $L$ is at most an $N$-bit integer.
2. Encode $N$ in unary either as $N$ zeros followed by a 1 or $N$ 1's followed by a 0.
3. Append $L$ as an $N$-bit number to this representation of $N$.

Section 2.7 describes yet another way to construct the same code. Section 2.14 shows a connection between this code and certain binary search trees.

Table 2.8 lists the first 18 gamma codes, where the $L$ part is in italics (see also Table 2.51 and the $C_1$ code of Table 2.16).

$$1 = 2^0 + 0 = 1 \qquad 10 = 2^3 + 2 = 0001\mathit{010}$$
$$2 = 2^1 + 0 = 01\mathit{0} \qquad 11 = 2^3 + 3 = 0001\mathit{011}$$
$$3 = 2^1 + 1 = 01\mathit{1} \qquad 12 = 2^3 + 4 = 0001\mathit{100}$$
$$4 = 2^2 + 0 = 001\mathit{00} \qquad 13 = 2^3 + 5 = 0001\mathit{101}$$
$$5 = 2^2 + 1 = 001\mathit{01} \qquad 14 = 2^3 + 6 = 0001\mathit{110}$$
$$6 = 2^2 + 2 = 001\mathit{10} \qquad 15 = 2^3 + 7 = 0001\mathit{111}$$
$$7 = 2^2 + 3 = 001\mathit{11} \qquad 16 = 2^4 + 0 = 00001\mathit{0000}$$
$$8 = 2^3 + 0 = 0001\mathit{000} \qquad 17 = 2^4 + 1 = 00001\mathit{0001}$$
$$9 = 2^3 + 1 = 0001\mathit{001} \qquad 18 = 2^4 + 2 = 00001\mathit{0010}$$

Peter Elias

Table 2.8: 18 Elias Gamma Codes.

In his 1975 paper, Elias describes two versions of the gamma code. The first version (titled $\gamma$) is encoded as follows:

1. Generate the binary representation $\beta(n)$ of $n$.
2. Denote the length $|\beta(n)|$ of $\beta(n)$ by $M$.
3. Generate the unary $u(M)$ representation of $M$ as $M - 1$ zeros followed by a 1.
4. Follow each bit of $\beta(n)$ by a bit of $u(M)$.
5. Drop the leftmost bit (the leftmost bit of $\beta(n)$ is always 1).

Thus, for $n = 13$ we prepare $\beta(13) = \overline{1}\overline{1}0\overline{1}$, so $M = 4$ and $u(4) = 0001$, resulting in $\overline{1}0\overline{1}0\overline{0}0\overline{1}1$. The final code is $\gamma(13) = 0\overline{1}0\overline{0}0\overline{1}1$.

The second version, dubbed $\gamma'$, moves the bits of $u(M)$ to the left. Thus $\gamma'(13) = 0001|\overline{1}0\overline{1}$. The gamma codes of Table 2.8 are Elias's $\gamma'$ codes. Both gamma versions are universal.

**Decoding** is also simple and is done in two steps:

1. Read zeros from the code until a 1 is encountered. Denote the number of zeros by $N$.

2. Read the next $N$ bits as an integer $L$. Compute $n = 2^N + L$.

It is easy to see that this code can be used to encode positive integers even in cases where the largest integer is not known in advance. Also, this code grows slowly (see Figure 2.34), so it is a good candidate for compressing integer data where small integers are common and large ones are rare.

**Elias delta code.** In his gamma code, Elias prepends the length of the code in unary ($\alpha$). In his next code, $\delta$ (delta), he prepends the length in binary ($\beta$). Thus, the Elias delta code, also for the positive integers, is slightly more complex to construct.

**Encoding** a positive integer $n$, is done in the following steps:

1. Write $n$ in binary. The leftmost (most-significant) bit will be a 1.

2. Count the bits, remove the leftmost bit of $n$, and prepend the count, in binary, to what is left of $n$ after its leftmost bit has been removed.

3. Subtract 1 from the count of step 2 and prepend that number of zeros to the code.

When these steps are applied to the integer 17, the results are: $17 = 10001_2$ (five bits). Remove the leftmost 1 and prepend $5 = 101_2$ yields $101|0001$. Three bits were added, so we prepend two zeros to obtain the delta code $00|101|0001$.

To compute the length of the delta code of $n$, we notice that step 1 generates (from Equation (1.1)) $M = 1 + \lfloor \log_2 n \rfloor$ bits. For simplicity, we omit the $\lfloor$ and $\rfloor$ and observe that

$$M = 1 + \log_2 n = \log_2 2 + \log_2 n = \log_2(2n).$$

The count of step 2 is $M$, whose length $C$ is therefore $C = 1 + \log_2 M = 1 + \log_2(\log_2(2n))$ bits. Step 2 therefore prepends $C$ bits and removes the leftmost bit of $n$. Step 3 prepends $C - 1 = \log_2 M = \log_2(\log_2(2n))$ zeros. The total length of the delta code is therefore the 3-part sum

$$\underbrace{\log_2(2n)}_{\text{step 1}} + \underbrace{[1 + \log_2\log_2(2n)] - 1}_{\text{step 2}} + \underbrace{\log_2\log_2(2n)}_{\text{step 3}} = 1 + \lfloor \log_2 n \rfloor + 2\lfloor \log_2\log_2(2n) \rfloor.$$

$$(2.2)$$

Figure 2.34 illustrates the length graphically. We show below that this code is ideal for data where the integer $n$ occurs with probability $1/[2n(\log_2(2n))^2]$.

An equivalent way to construct the delta code employs the gamma code:

1. Find the largest integer $N$ such that $2^N \geq n < 2^{N+1}$ and write $n = 2^N + L$. Notice that $L$ is at most an $N$-bit integer.

2. Encode $N + 1$ with the Elias gamma code.

3. Append the binary value of $L$, as an $N$-bit integer, to the result of step 2.

When these steps are applied to $n = 17$, the results are: $17 = 2^N + L = 2^4 + 1$. The gamma code of $N + 1 = 5$ is 00101, and appending $L = 0001$ to this yields 00101|0001.

Table 2.9 lists the first 18 delta codes, where the $L$ part is in italics. See also the related code $C_3$ of Table 2.16, which has the same length.

| | |
|---|---|
| $1 = 2^0 + 0 \to \|L\| = 0 \to 1$ | $10 = 2^3 + 2 \to \|L\| = 3 \to 00100\mathit{010}$ |
| $2 = 2^1 + 0 \to \|L\| = 1 \to 010\mathit{0}$ | $11 = 2^3 + 3 \to \|L\| = 3 \to 00100\mathit{011}$ |
| $3 = 2^1 + 1 \to \|L\| = 1 \to 010\mathit{1}$ | $12 = 2^3 + 4 \to \|L\| = 3 \to 00100\mathit{100}$ |
| $4 = 2^2 + 0 \to \|L\| = 2 \to 011\mathit{00}$ | $13 = 2^3 + 5 \to \|L\| = 3 \to 00100\mathit{101}$ |
| $5 = 2^2 + 1 \to \|L\| = 2 \to 011\mathit{01}$ | $14 = 2^3 + 6 \to \|L\| = 3 \to 00100\mathit{110}$ |
| $6 = 2^2 + 2 \to \|L\| = 2 \to 011\mathit{10}$ | $15 = 2^3 + 7 \to \|L\| = 3 \to 00100\mathit{111}$ |
| $7 = 2^2 + 3 \to \|L\| = 2 \to 011\mathit{11}$ | $16 = 2^4 + 0 \to \|L\| = 4 \to 00101\mathit{0000}$ |
| $8 = 2^3 + 0 \to \|L\| = 3 \to 00100\mathit{000}$ | $17 = 2^4 + 1 \to \|L\| = 4 \to 00101\mathit{0001}$ |
| $9 = 2^3 + 1 \to \|L\| = 3 \to 00100\mathit{001}$ | $18 = 2^4 + 2 \to \|L\| = 4 \to 00101\mathit{0010}$ |

Table 2.9: 18 Elias Delta Codes.

**Decoding** is done in the following steps:

1. Read bits from the code until you can decode an Elias gamma code. Call the decoded result $M + 1$. This is done in the following substeps:

1.1 Count the leading zeros of the code and denote the count by $C$.

1.2 Examine the leftmost $2C + 1$ bits ($C$ zeros, followed by a single 1, followed by $C$ more bits). This is the decoded gamma code $M + 1$.

2. Read the next $M$ bits. Call this number $L$.

3. The decoded integer is $2^M + L$.

In the case of $n = 17$, the delta code is 001010001. We skip two zeros, so $C = 2$. The value of the leftmost $2C + 1 = 5$ bits is 00101 = 5, so $M + 1 = 5$. We read the next $M = 4$ bits 0001, and end up with the decoded value $2^M + L = 2^4 + 1 = 17$.

To better understand the application and performance of these codes, we need to identify the type of data it compresses best. Given a set of symbols $a_i$, where each symbol occurs in the data with probability $P_i$ and the length of its code is $l_i$ bits, the average code length is the sum $\sum P_i l_i$ and the entropy (the smallest number of bits required to represent the symbols) is $\sum [-P_i \log_2 P_i]$. The redundancy (Equation (1.2)) is the difference $\sum_i P_i l_i - \sum_i [-P_i \log_2 P_i]$ and we are looking for probabilites $P_i$ that will minimize this difference.

In the case of the gamma code, $l_i = 1 + 2 \log_2 i$. If we select symbol probabilities $P_i = 1/(2i^2)$ (a power law distribution of probabilities, where the first 10 values are 0.5, 0.125, 0.0556, 0.03125, 0.02, 0.01389, 0.0102, 0.0078, 0.00617, and 0.005, see also Table 2.17), both the average code length and the entropy become the identical sums

$$\sum_i \frac{1 + 2 \log i}{2i^2},$$

indicating that the gamma code is asymptotically optimal for this type of data. A power law distribution of values is dominated by just a few symbols and especially by the first. Such a distribution is very skewed and is therefore handled very well by the gamma code which starts very short. In an exponential distribution, in contrast, the small values have similar probabilities, which is why data with this type of statistical distribution is compressed better by a Rice code (Section 2.24).

In the case of the delta code, $l_i = 1 + \log i + 2 \log \log(2i)$. If we select symbol probabilities $P_i = 1/[2i(\log(2i))^2]$ (where the first five values are 0.5, 0.0625, 0.025, 0.0139, and 0.009), both the average code length and the entropy become the identical sums

$$\sum_i \frac{\log 2 + \log i + 2 \log \log(2i)}{2i(\log(2i))^2},$$

indicating that the redundancy is zero and the delta code is therefore asymptotically optimal for this type of data.

Section 2.14 shows a connection between a variant of the delta code and certain binary search trees.

> The phrase "working mother" is redundant.
> —Jane Sellman

**The Elias omega code.** Unlike the previous Elias codes, the omega code uses itself recursively to encode the prefix $M$, which is why it is sometimes referred to as a recursive Elias code. The main idea is to prepend the length of $n$ to $n$ as a group of bits that starts with a 1, then prepend the length of the length, as another group, to the result, and continue prepending lengths until the last length is 2 or 3 (and therefore fits in two bits). In order to distinguish between a length group and the last, rightmost group (of $n$ itself), the latter is followed by a delimiter of 0, while each length group starts with a 1.

**Encoding** a positive integer $n$ is done recursively in the following steps:

1. Initialize the code-so-far to 0.

2. If the number to be encoded is 1, stop; otherwise, prepend the binary representation of $n$ to the code-so-far. Assume that we have prepended $L$ bits.

3. Repeat step 2, with the binary representation of $L - 1$ instead of $n$.

The integer 17 is therefore encoded by (1) a single 0, (2) prepended by the 5-bit binary value 10001, (3) prepended by the 3-bit value of $5 - 1 = 100_2$, and (4) prepended by the 2-bit value of $3 - 1 = 10_2$. The result is 10|100|10001|0.

Table 2.10 lists the first 18 omega codes (see also Table 2.13). Note that $n = 1$ is handled as a special case.

| | | | |
|---|---|---|---|
| 1 | 0 | 10 | 11 1010 0 |
| 2 | 10 0 | 11 | 11 1011 0 |
| 3 | 11 0 | 12 | 11 1100 0 |
| 4 | 10 100 0 | 13 | 11 1101 0 |
| 5 | 10 101 0 | 14 | 11 1110 0 |
| 6 | 10 110 0 | 15 | 11 1111 0 |
| 7 | 10 111 0 | 16 | 10 100 10000 0 |
| 8 | 11 1000 0 | 17 | 10 100 10001 0 |
| 9 | 11 1001 0 | 18 | 10 100 10010 0 |

Table 2.10: 18 Elias Omega Codes.

**Decoding** is done in several nonrecursive steps where each step reads a group of bits from the code. A group that starts with a zero signals the end of decoding.

1. Initialize $n$ to 1.

2. Read the next bit. If it is 0, stop. Otherwise read $n$ more bits, assign the group of $n + 1$ bits to $n$, and repeat this step.

Some readers may find it easier to understand these steps rephrased as follows.

1. Read the first group, which will either be a single 0, or a 1 followed by $n$ more digits. If the group is a 0, the value of the integer is 1; if the group starts with a 1, then $n$ becomes the value of the group interpreted as a binary number.

2. Read each successive group; it will either be a single 0, or a 1 followed by $n$ more digits. If the group is a 0, the value of the integer is $n$; if it starts with a 1, then $n$ becomes the value of the group interpreted as a binary number.

**Example**. Decode 10|100|10001|0. The decoder initializes $n = 1$ and reads the first bit. It is a 1, so it reads $n = 1$ more bit (0) and assigns $n = 10_2 = 2$. It reads the next bit. It is a 1, so it reads $n = 2$ more bits (00) and assigns the group 100 to $n$. It reads the next bit. It is a 1, so it reads four more bits (0001) and assigns the group 10001 to $n$. The next bit read is 0, indicating the end of decoding.

The omega code is constructed recursively, which is why its length $|\omega(n)|$ can also be computed recursively. We define the quantity $l^k(n)$ recursively by $l^1(n) = \lfloor \log_2 n \rfloor$

and $l^{i+1}(n) = l^1(l^i(n))$. Equation (1.1) tells us that $|\beta(n)| = l^1(n) + 1$ (where $\beta$ is the standard binary representation), and this implies that the length of the omega code is given by the sum

$$|\omega(n)| = \sum_{i=1}^{k} \beta(l^{k-i}(n)) + 1 = 1 + \sum_{i=1}^{k} (l^i(n) + 1),$$

where the sum stops at the $k$ that satisfies $l^k(n) = 1$. From this, Elias concludes that the length satisfies $|\omega(n)| \leq 1 + \frac{5}{2}\lfloor \log_2 n \rfloor$.

A quick glance at any table of these codes shows that their lengths fluctuate. In general, the length increases slowly as $n$ increases, but when a new length group is added, which happens when $n = 2^{2^k}$ for any positive integer $k$, the length of the code increases suddenly by several bits. For $k$ values of 1, 2, 3, and 4, this happens when $n$ reaches 4, 16, 256, and 65,536. Because the groups of lengths are of the form "length," "log(length)," "log(log(length))," and so on, the omega code is sometimes referred to as a logarithmic-ramp code.

Table 2.11 compares the length of the gamma, delta, and omega codes. It shows that the delta code is asymptotically best, but if the data consists mostly of small numbers (less than 8) and there are only a few large integers, then the gamma code performs better.

| Values | Gamma | Delta | Omega |
|--------|-------|-------|-------|
| 1 | 1 | 1 | 2 |
| 2 | 3 | 4 | 3 |
| 3 | 3 | 4 | 4 |
| 4 | 5 | 5 | 4 |
| 5–7 | 5 | 5 | 5 |
| 8–15 | 7 | 8 | 6–7 |
| 16–31 | 9 | 9 | 7–8 |
| 32–63 | 11 | 10 | 8–10 |
| 64–88 | 13 | 11 | 10 |
| 100 | 13 | 11 | 11 |
| 1000 | 19 | 16 | 16 |
| $10^4$ | 27 | 20 | 20 |
| $10^5$ | 33 | 25 | 25 |
| $10^5$ | 39 | 28 | 30 |

Table 2.11: Lengths of Three Elias Codes.

Beware of bugs in the above code; I have only proved it correct, not tried it.
—Donald Knuth

## 2.5 Levenstein Code

This little-known code for the nonnegative integers was conceived in 1968 by Vladimir Levenshtein [Levenstein 06]. Both encoding and decoding are multistep processes.

**Encoding.** The Levenstein code of zero is a single 0. To code a positive number $n$, perform the following:

1. Set the count variable $C$ to 1. Initialize the code-so-far to the empty string.
2. Take the binary value of $n$ without its leading 1 and prepend it to the code-so-far.
3. Let $M$ be the number of bits prepended in step 2.
4. If $M \neq 0$, increment $C$ by 1, go to and execute step 2 with $M$ instead of $n$.
5. If $M = 0$, prepend $C$ 1's followed by a 0 to the code-so-far and stop.

| $n$ | Levenstein code | $n$ | Levenstein code |
|---|---|---|---|
| 0 | 0 | 9 | 1110  1  001 |
| 1 | 10 | 10 | 1110  1  010 |
| 2 | 110  0 | 11 | 1110  1  011 |
| 3 | 110  1 | 12 | 1110  1  100 |
| 4 | 1110  0  00 | 13 | 1110  1  101 |
| 5 | 1110  0  01 | 14 | 1110  1  110 |
| 6 | 1110  0  10 | 15 | 1110  1  111 |
| 7 | 1110  0  11 | 16 | 11110  0  00   0000 |
| 8 | 1110  1  000 | 17 | 11110  0  00   0001 |

Table 2.12: 18 Levenstein Codes.

Table 2.12 lists some of these codes. Spaces have been inserted to indicate the individual parts of each code. As an exercise, the reader may verify that the Levenstein codes for 18 and 19 are 11110|0|00|0010 and 11110|0|00|0011, respectively.

**Decoding** is done as follows:

1. Set count $C$ to the number of consecutive 1's preceding the first 0.
2. If $C = 0$, the decoded value is zero, stop.
3. Set $N = 1$, and repeat step 4 $(C - 1)$ times.
4. Read $N$ bits, prepend a 1, and assign the resulting bitstring to $N$ (thereby erasing the previous value of $N$). The string assigned to $N$ in the last iteration is the decoded value.

The Levenstein code of the positive integer $n$ is always one bit longer than the Elias omega code of $n$.

# 2.6 Even–Rodeh Code

The principle behind the Elias omega code occurred independently to S. Even and M. Rodeh [Even and Rodeh 70]. Their code is similar to the omega code, the main difference being that lengths are prepended until a 3-bit length is reached and becomes the leftmost group of the code. For example, the Even–Rodeh code of 2761 is 100|1100|101011001001|0 (4, 12, 2761, and 0).

The authors prove, by induction on $n$, that every nonnegative integer can be encoded in this way. They also show that the length $l(n)$ of their code satisfies $l(n) \leq 4 + 2L(n)$ where $L(n)$ is the length of the binary representation (beta code) of $n$, Equation (1.1).

The developers show how this code can be used as a comma to separate variable-length symbols in a string. Given a string of symbols $a_i$, precede each $a_i$ with the Even–Rodeh code $R$ of its length $l_i$. The codes and symbols can then be concatenated into a single string

$$R(l_1)a_1 R(l_2)a_2 \ldots R(l_m)a_m 000$$

that can be uniquely separated into the individual symbols because the codes act as separators (commas). Three zeros act as the string delimiter.

The developers also prove that the extra length of the codes shrinks asymptotically to zero as the string becomes longer. For strings of length $2^{10}$ bits, the overhead is less than 2%. Table 2.13 (after [Fenwick 96]) lists several omega and Even–Rodeh codes.

| $n$ | Omega | Even–Rodeh |
|---|---|---|
| 0 | — | 000 |
| 1 | 0 | 001 |
| 2 | 10 0 | 010 |
| 3 | 11 0 | 011 |
| 4 | 10 100 0 | 100 0 |
| 7 | 10 111 0 | 111 0 |
| 8 | 11 1000 0 | 100 1000 0 |
| 15 | 11 1111 0 | 100 1111 0 |
| 16 | 10 100 10000 0 | 101 10000 0 |
| 32 | 10 101 100000 0 | 110 100000 0 |
| 100 | 10 110 1100100 0 | 111 1100100 0 |
| 1000 | 10 110 1100100 0 | 110 1100100 0 |

| Values | $\omega$ | ER |
|---|---|---|
| 1 | 1 | 3 |
| 2–3 | 3 | 3 |
| 4–7 | 6 | 4 |
| 8–15 | 7 | 8 |
| 16–31 | 11 | 9 |
| 32–63 | 12 | 10 |
| 64–127 | 13 | 11 |
| 128–255 | 14 | 17 |
| 256–512 | 21 | 18 |

Table 2.13: Several Omega and Even–Rodeh Codes.     Table 2.14: Different Lengths.

Table 2.14 (after [Fenwick 96]) illustrates the different points where the lengths of these codes increase. The length of the omega code increases when $n$ changes from the form $2^m - 1$ (where the code is shortest relative to $n$) to $2^m$ (where it is longest). The Even–Rodeh code behaves similarly, but may be slightly shorter or longer for various intervals of $n$.

# 2.7 Punctured Elias Codes

The punctured Elias codes for the integers were designed by Peter Fenwick in an attempt to improve the performance of the Burrows–Wheeler transform [Burrows and Wheeler 94]. The codes are described in the excellent, 15-page technical report [Fenwick 96]. The term "punctured" comes from the field of error-control codes. Often, a codeword for error-detection or error-correction consists of the original data bits plus a number of check bits. If some check bits are removed, to shorten the codeword, the resulting code is referred to as punctured.

We start with the Elias gamma code. Section 2.4 describes two ways to construct this code, and here we consider a third approach to the same construction. Write the binary value of $n$, reverse its bits so its rightmost bit is now a 1, and prepend flags for the bits of $n$. For each bit of $n$, create a flag of 0, except for the last (rightmost) bit (which is a 1), whose flag is 1. Prepend the flags to the reversed $n$ and remove the rightmost bit. Thus, $13 = 1101_2$ is reversed to yield 1011. The four flags 0001 are prepended and the rightmost bit of 1011 is removed to yield the final gamma code 0001|101.

The punctured code eliminates the flags for zeros and is constructed as follows. Write the binary value of $n$, reverse its bits, and prepend flags to indicate the number of 1's in $n$. For each bit of 1 in $n$ we prepare a flag of 1, and terminate the flags with a single 0. Thus, $13 = 1101_2$ is reversed to 1011. It has three 1's, so the flags 1110 are prepended to yield 1110|1011. We call this punctured code $P1$ and notice that it starts with a 1 (there is at least one flag, except for the $P1$ code of $n = 0$) and also ends with a 1 (because the original $n$, whose MSB is a 1, has been reversed). We can therefore construct another punctured code $P2$ such that $P2(n)$ equals $P1(n+1)$ with its most-significant 1 removed.

Table 2.15 lists examples of $P1$ and $P2$. The feature that strikes the reader most is that the codes are generally getting longer as $n$ increases, but are also getting shorter from time to time, for $n$ values that are 1 less than a power of 2. For small values of $n$, these codes are often a shade longer than the gamma code, but for large values they average about $1.5 \log_2 n$ bits, shorter than the $2\lfloor \log_2 n \rfloor + 1$ bits of the Elias gamma code.

One design consideration for these codes was their expected behavior when applied to data with a skewed distribution and more smaller values than larger values. Smaller binary integers tend to have fewer 1's, so it was hoped that the punctures would reduce the average code length below $1.5 \log_2 n$ bits. Later work by Fenwick showed that this hope did not materialize.

> And as they carried him away
> Our punctured hero was heard to say,
> When in this war you venture out,
> Best never do it dressed as a Kraut!
>
> —Stephen Ambrose, *Band of Brothers*

| $n$ | $P1$ | $P2$ | $n$ | $P1$ | $P2$ |
|---|---|---|---|---|---|
| 0 | 0 | 01 | 11 | 11101101 | 100011 |
| 1 | 101 | 001 | 12 | 1100011 | 1101011 |
| 2 | 1001 | 1011 | 10 | 11101011 | 1100111 |
| 3 | 11011 | 0001 | 14 | 11100111 | 11101111 |
| 4 | 10001 | 10101 | 15 | 111101111 | 000001 |
| 5 | 110101 | 10011 | 16 | 1000001 | 1010001 |
| 6 | 110011 | 110111 | ... | | |
| 7 | 1110111 | 00001 | 31 | 11111011111 | 0000001 |
| 8 | 100001 | 101001 | 32 | 10000001 | 10100001 |
| 9 | 1101001 | 100101 | 33 | 110100001 | 10010001 |
| 10 | 1100101 | 1101101 | | | |

Table 2.15: Two Punctured Elias Codes.

# 2.8 Other Prefix Codes

Four prefix codes, $C_1$ through $C_4$, are presented in this section. We denote by $B(n)$ the binary representation of the integer $n$ (the beta code). Thus, $|B(n)|$ is the length, in bits, of this representation. We also use $\overline{B}(n)$ to denote $B(n)$ without its most significant bit (which is always 1).

Code $C_1$ consists of two parts. To code the positive integer $n$, we first generate the unary code of $|B(n)|$ (the size of the binary representation of $n$), then append $\overline{B}(n)$ to it. An example is $n = 16 = 10000_2$. The size of B(16) is 5, so we start with the unary code 11110 (or 00001) and append $\overline{B}(16) = 0000$. Thus, the complete code is 11110|0000 (or 00001|0000). Another example is $n = 5 = 101_2$ whose code is 110|01. The length of $C_1(n)$ is $2\lfloor \log_2 n \rfloor + 1$ bits. Notice that this code is identical to the general unary code $(0, 1, \infty)$ and is closely related to the Elias gamma code.

Code $C_2$ is a rearrangement of $C_1$ where each of the $1 + \lfloor \log_2 n \rfloor$ bits of the first part (the unary code) of $C_1$ is followed by one of the bits of the second part. Thus, code $C_2(16) = 101010100$ and $C_2(5) = 10110$.

Code $C_3$ starts with $|B(n)|$ coded in $C_2$, followed by $\overline{B}(n)$. Thus, 16 is coded as $C_2(5) = 10110$ followed by $\overline{B}(16) = 0000$, and 5 is coded as code $C_2(3) = 110$ followed by $\overline{B}(5) = 01$. The length of $C_3(n)$ is $1 + \lfloor \log_2 n \rfloor + 2\lfloor \log_2(1 + \lfloor \log_2 n \rfloor) \rfloor$ (same as the length of the Elias delta code, Equation (2.2)).

Code $C_4$ consists of several parts. We start with $B(n)$. To its left we prepend the binary representation of $|B(n)| - 1$ (one less than the length of $n$). This continues recursively, until a 2-bit number is written. A 0 is then appended to the right of the entire code, to make it uniquely decodable. To encode 16, we start with 10000, prepend $|B(16)| - 1 = 4 = 100_2$ to the left, then prepend $|B(4)| - 1 = 2 = 10_2$ to the left of that, and finally append a 0 on the right. The result is 10|100|10000|0. To encode 5, we start with 101, prepend $|B(5)| - 1 = 2 = 10_2$ to the left, and append a 0 on the right. The result is 10|101|0. Comparing with Table 2.10 shows that $C_4$ is the omega code.

(The 0 on the right make the code uniquely decodable because each part of $C_4$ is the standard binary code of some integer, so it starts with a 1. A start bit of 0 is therefore a signal to the decoder that this is the last part of the code.)

| $n$ | $B(n)$ | $\overline{B}(n)$ | $C_1$ | $C_2$ | $C_3$ | $C_4$ |
|---|---|---|---|---|---|---|
| 1 | 1 | | 0\| | 0 | 0\| | 0 |
| 2 | 10 | 0 | 10\|0 | 100 | 100\|0 | 10\|0 |
| 3 | 11 | 1 | 10\|1 | 110 | 100\|1 | 11\|0 |
| 4 | 100 | 00 | 110\|00 | 10100 | 110\|00 | 10\|100\|0 |
| 5 | 101 | 01 | 110\|01 | 10110 | 110\|01 | 10\|101\|0 |
| 6 | 110 | 10 | 110\|10 | 11100 | 110\|10 | 10\|110\|0 |
| 7 | 111 | 11 | 110\|11 | 11110 | 110\|11 | 10\|111\|0 |
| 8 | 1000 | 000 | 1110\|000 | 1010100 | 10100\|000 | 11\|1000\|0 |
| 9 | 1001 | 001 | 1110\|001 | 1010110 | 10100\|001 | 11\|1001\|0 |
| 10 | 1010 | 010 | 1110\|010 | 1011100 | 10100\|010 | 11\|1010\|0 |
| 11 | 1011 | 011 | 1110\|011 | 1011110 | 10100\|011 | 11\|1011\|0 |
| 12 | 1100 | 100 | 1110\|100 | 1110100 | 10100\|100 | 11\|1100\|0 |
| 13 | 1101 | 101 | 1110\|101 | 1110110 | 10100\|101 | 11\|1101\|0 |
| 14 | 1110 | 110 | 1110\|110 | 1111100 | 10100\|110 | 11\|1110\|0 |
| 15 | 1111 | 111 | 1110\|111 | 1111110 | 10100\|111 | 11\|1111\|0 |
| 16 | 10000 | 0000 | 11110\|0000 | 101010100 | 10110\|0000 | 10\|100\|10000\|0 |
| 31 | 11111 | 1111 | 11110\|1111 | 111111110 | 10110\|1111 | 10\|100\|11111\|0 |
| 32 | 100000 | 00000 | 111110\|00000 | 10101010100 | 11100\|00000 | 10\|101\|100000\|0 |
| 63 | 111111 | 11111 | 111110\|11111 | 11111111110 | 11100\|11111 | 10\|101\|111111\|0 |
| 64 | 1000000 | 000000 | 1111110\|000000 | 1010101010100 | 11110\|000000 | 10\|110\|1000000\|0 |
| 127 | 1111111 | 111111 | 1111110\|111111 | 1111111111110 | 11110\|111111 | 10\|110\|1111111\|0 |
| 128 | 10000000 | 0000000 | 11111110\|0000000 | 101010101010100 | 1010100\|0000000 | 10\|111\|10000000\|0 |
| 255 | 11111111 | 1111111 | 11111110\|1111111 | 111111111111110 | 1010100\|1111111 | 10\|111\|11111111\|0 |

Table 2.16: Some Prefix Codes.

| $n$ | Unary | $C_1$ | $C_3$ |
|---|---|---|---|
| 1 | 0.5 | 0.5000000 | |
| 2 | 0.25 | 0.1250000 | 0.2500000 |
| 3 | 0.125 | 0.0555556 | 0.0663454 |
| 4 | 0.0625 | 0.0312500 | 0.0312500 |
| 5 | 0.03125 | 0.0200000 | 0.0185482 |
| 6 | 0.015625 | 0.0138889 | 0.0124713 |
| 7 | 0.0078125 | 0.0102041 | 0.0090631 |
| 8 | 0.00390625 | 0.0078125 | 0.0069444 |

Table 2.17: Ideal Probabilities of Eight Integers for Three Codes.

Table 2.16 shows examples of the four codes above, as well as $B(n)$ and $\overline{B}(n)$. The lengths of the four codes shown in the table increase as $\log_2 n$, in contrast with the length of the unary code, which increases as $n$. These codes are therefore good choices in cases where the index constant of integers is with prohibitation that certain certain conditions. Specifically, the length $L$ of the unary code of $n$ is $L = n = \log_2 2^n$, so it is ideal for the case where $P(n) = 2^{-L} = 2^{-n}$. The length of code $C_1(n)$ is $L = 1 + 2\lfloor \log_2 n \rfloor = \log_2 2 + \log_2 n^2 = \log_2(2n^2)$, so it is ideal for the case where

$$P(n) = 2^{-L} = \frac{1}{2n^2}.$$

The length of code $C_3(n)$ is

$$L = 1 + \lfloor \log_2 n \rfloor + 2\lfloor \log_2(1 + \lfloor \log_2 n \rfloor) \rfloor = \log_2 2 + 2\lfloor \log \log_2 2n \rfloor + \lfloor \log_2 n \rfloor,$$

so it is ideal for the case where

$$P(n) = 2^{-L} = \frac{1}{2n(\log_2 n)^2}.$$

Table 2.17 shows the ideal probabilities that the first eight positive integers should have for the unary, $C_1$, and $C_3$ codes to be used.

More prefix codes for the positive integers, appropriate for special applications, may be designed by the following general approach. Select positive integers $v_i$ and combine them in a list $V$ (which may be finite or infinite according to needs). The code of the positive integer $n$ is prepared in the following steps:

1. Find $k$ such that

$$\sum_{i=1}^{k-1} v_i < n \le \sum_{i=1}^{k} v_i.$$

2. Compute the difference

$$d = n - \left[\sum_{i=1}^{k-1} v_i\right] - 1.$$

The largest value of $n$ is $\sum_{i}^{k} v_i$, so the largest value of $d$ is $\sum_{i}^{k} v_i - \left[\sum_{1}^{k-1} v_i\right] - 1 = v_k - 1$, a number that can be written in $\lceil \log_2 v_k \rceil$ bits. The number $d$ is encoded, using the standard binary code, either in this number of bits, or if $d < 2^{\lceil \log_2 v_k \rceil} - v_k$, it is encoded in $\lfloor \log_2 v_k \rfloor$ bits.

3. Encode $n$ in two parts. Start with $k$ encoded in some prefix code, and concatenate the binary code of $d$. Since $k$ is coded in a prefix code, any decoder would know how many bits to read for $k$. After reading and decoding $k$, the decoder can compute the value $2^{\lceil \log_2 v_k \rceil} - v_k$ which tells it how many bits to read for $d$.

A simple example is the infinite sequence $V = (1, 2, 4, 8, \ldots, 2^{i-1}, \ldots)$ with $k$ coded in unary. The integer $n = 10$ satisfies

$$\sum_{i=1}^{3} v_i < 10 \le \sum_{i=1}^{4} v_i,$$

so $k = 4$ (with unary code 1110) and $d = 10 - \left[\sum_{i=1}^{3} v_i\right] - 1 = 2$. Our values $v_i$ are powers of 2, so $\log_2 v_i$ is an integer and $2^{\log_2 v_k}$ equals $v_i$. Thus, the length of $d$ in our example is $\log_2 v_i = \log_2 8 = 3$ and the code of 10 is 1110|010.

## 2.9 Ternary Comma Code

Binary (base 2) numbers are based on the two bits 0 and 1. Similarly, ternary (base 3) numbers are based on the three digits (trits) 0, 1, and 2. Each trit can be encoded in two bits, but two bits can have four values. Thus, it makes sense to work with a ternary number system where each trit is represented by two bits and in addition to the three trits there is a fourth symbol, a comma ($c$). Once we include the $c$, it becomes easy to construct the ternary comma code for the integers. The comma code of $n$ is simply the ternary representation of $n - 1$ followed by a $c$. Thus, the comma code of 8 is $21c$ (because $7 = 2 \cdot 3 + 1$) and the comma code of 18 is $122c$ (because $17 = 1 \cdot 9 + 2 \cdot 3 + 2$).

Table 2.18 (after [Fenwick 96]) lists several ternary comma codes (the columns labeled L are the length of the code, in bits). These codes start long (longer than most of the other codes described here) but grow slowly. Thus, they are suitable for applications where large integers are common. These codes are also easy to encode and decode and their principal downside is the comma symbol (signalling the end of a code) that requires two bits. This inefficiency is not serious, but becomes more so for comma codes based on larger number bases. In a base-15 comma code, for example, each of the 15 digits requires four bits and the comma is also a 4-bit pattern. Each code ends with a 4-bit comma, instead of with the theoretical minimum of one bit, and this feature renders such codes inefficient. (However, the overall redundancy per symbol decreases for large number bases. In a base-7 system, one of eight symbols is sacrificed for the comma, while in a base 15 it is one of 16 symbols.)

| Value | Code | L | Value | Code | L |
|---|---|---|---|---|---|
| 0 | c | 2 | 11 | 101c | 8 |
| 1 | 0c | 4 | 12 | 102c | 8 |
| 2 | 1c | 4 | 13 | 110c | 8 |
| 3 | 2c | 4 | 14 | 111c | 8 |
| 4 | 10c | 6 | 15 | 112c | 8 |
| 5 | 11c | 6 | 16 | 120c | 8 |
| 6 | 12c | 6 | 17 | 121c | 8 |
| 7 | 20c | 6 | 18 | 122c | 8 |
| 8 | 21c | 6 | 19 | 200c | 8 |
| 9 | 22c | 6 | 20 | 201c | 8 |
| ... | | | ... | | |
| 64 | 2100c | 10 | 1,000 | 1101000c | 16 |
| 128 | 11201c | 12 | 3,000 | 11010002c | 18 |
| 256 | 100110c | 14 | 10,000 | 111201100c | 20 |
| 512 | 200221c | 14 | 65,536 | 10022220020c | 24 |

Table 2.18: Ternary Comma Codes and Their Lengths.

The ideas above are simple and easy to implement, but they make sense only for long bitstrings. In data compression, as well as in many other applications, we normally need short codes, ranging in size from two to perhaps 12 bits. Because they are used for compression, such codes have to be self-delimiting without adding any extra bits. The solution is to design sets of prefix codes, codes that have the prefix property. Among these codes, the ones that yield the best performance are the universal codes.

In 1840, Thomas Fowler, a self-taught English mathematician and inventor, created a unique ternary calculating machine. Until recently, all detail of this machine was lost. A research project begun in 1997 uncovered sufficient information to enable the recreation of a physical concept model of Fowler's machine. The next step is to create a historically accurate replica.

—[Glusker et al 05]

# 2.10 Location Based Encoding (LBE)

Location based encoding (LBE) is a simple method for assigning variable-length codes to a set of symbols, not necessarily integers. The originators of this method are P. S. Chitaranjan, Arun Shankar, and K. Niyant [LBE 07]. The LBE encoder is essentially a two-pass algorithm. Given a data file to be compressed, the encoder starts by reading the file and counting symbol frequencies. Following the first pass, it (1) sorts the symbols in descending order of their probabilities, (2) stores them in a special order in a matrix, and (3) assigns them variable-length codes. In between the passes, the encoder writes the symbols, in their order in the matrix, on the compressed stream, for the use of the decoder.

The second pass simply reads the input file again symbol by symbol and replaces each symbol with its code. The decoder starts by reading the sorted sequence of symbols from the compressed stream and constructing the codes in lockstep with the encoder. The decoder then reads the codes off its input and replaces each code with the original symbol. Thus, decoding is fast.

The main idea is to assign short codes to the common symbols by placing the symbols in a matrix in a special diagonal order, vaguely reminiscent of the zigzag order used by JPEG, such that high-probability symbols are concentrated at the top-left corner of the matrix. This is illustrated in Figure 2.19a, where the numbers 1, 2, 3, ... indicate the most-common symbol, the second most-common one, and so on. Each symbol is assigned a variable-length code that indicates its position (row and column) in the matrix. Thus, the code of 1 (the most-common symbol) is 11 (row 1 column 1), the code of 6 is 0011 (row 3 column 1), and the code of 7 is 10001 (row 1 column 4). Each code has two 1's, which makes it trivial for the decoder to read and identify the codes. The length of a code depends on the diagonal (shown in gray in the figure), and the figure shows codes with lengths (also indicated in gray) from two to eight bits.

We know that Huffman coding fails for very small alphabets. For a 2-symbol alphabet, the Huffman algorithm assigns 1-bit codes to the two symbols regardless of their probabilities, so no compression is achieved. Given the same alphabet, LBE assigns the two codes 11 and 101 to the symbols, so the average length of its codes is between 2 and

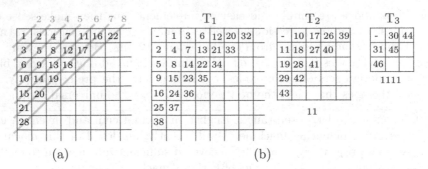

Figure 2.19: LBE Code Tables.

3 (the shorter code is assigned to the most-common symbol, which is why the average code length is in the interval $(2, 2.5]$, still bad). Thus, LBE performs even worse in this case. For alphabets with small numbers of symbols, such as three or four, LBE performs better, but still not as good as Huffman.

The ASCII code assigns 7-bit codes to 128 symbols, and this should be compared with the average code length achievable by LBE. LBE assigns codes of seven or fewer bits to the 21 most-common symbols. Thus, if the alphabet consists of 21 or fewer symbols, replacing them with LBE codes generates compression. For larger alphabets, longer codes are needed, which may cause expansion (i.e., compression where the average code length is greater than seven bits).

It is possible to improve LBE's performance somewhat by constructing several tables, as illustrated in Figure 2.19b. Each code is preceded by its table number such that the codes of table $n$ are preceded by $(2n - 2)$ 1's. The codes of table 1 are not preceded by any 1's, those in table 2 are preceded by 11, those of table 3 are preceded by 1111, and so on. Because of the presence of pairs of 1's, position $(1, 1)$ of a table cannot be used and the first symbol of a table is placed at position $(1, 2)$. The codes are placed in table 1 until a code can become shorter if placed in table 2. This happens with the 10th most-common symbol. Placing it in table 1 would have assigned it the code 100001, whereas placing it in table 2 (which is empty so far) assigns it the shorter code 11|101. The next example is the 12th symbol. Placing it in table 2 would have assigned it code 11|1001, but placing it in table 1 gives it the same-length code 100001. With this scheme it is possible to assign short codes (seven bits or fewer) to the 31 most-common symbols.

Notice that there is no need to actually construct any tables. Both encoder and decoder can determine the codes one by one simply by keeping track of the next available position in each table. Once a code has been computed, it can be stored with other codes in an array or another convenient data structure.

The decoder reads pairs of consecutive 1's until it reaches a 0 or a single 1. This provides the table number, and the rest of the code is read until two 1's (not necessarily consecutive) are found.

The method is simple but not very efficient, especially because the compressed stream has to start with a list of the symbols in ascending order of probabilities. The main advantage of LBE is simple implementation, simple encoding, and fast decoding.

> Die Mathematiker sind eine Art Franzosen: redet man zu ihnen, so uebersetzen sie es
> in ihre Sprache, und dann ist es also bald ganz etwas anderes.
> (The mathematicians are a sort of Frenchmen: when you talk to them, they imme-
> ~~diately translate it into their own language, and right away it is something entirely~~
> different.)
>
> —Johann Wolfgang von Goethe

## 2.11 Stout Codes

In his 1980 short paper [Stout 80], Quentin Stout introduced two families $R_l$ and $S_l$ of recursive, variable-length codes for the integers, similar to and more general than Elias omega and Even–Rodeh codes. The Stout codes are universal and asymptotically optimal. Before reading ahead, the reader is reminded that the length $L$ of the integer $n$ is given by $L = 1 + \lfloor \log_2 n \rfloor$ (Equation (1.1)).

The two families of codes depend on the choice of an integer parameter $l$. Once a value (greater than or equal to 2) for $l$ has been selected, a codeword in the first family consists of a prefix $R_l(n)$ and a suffix $0n$.

The suffix is the binary value of $n$ in $L$ bits, preceded by a single 0 separator. The prefix $R_l(n)$ consists of length groups. It starts with the length $L$ of $n$, to which is prepended the length $L_1$ of $L$, then the length $L_2$ of $L_1$, and so on until a length $L_i$ is reached that is short enough to fit in an $l$-bit group. Notice that the length groups (except perhaps the leftmost one) start with 1. Thus, the single 0 separator indicates the end of the length groups.

As an example, we select $l = 2$ and encode a 985-bit integer $n$ whose precise value is irrelevant. The suffix is the 986-bit string $0n$ and the prefix starts with the group $L = 985 = 1111011001_2$. The length $L_1$ of this group is $10_{10} = 1010_2$, the length $L_2$ of the second group is $4 = 100_2$, and the final length $L_3$ is $3 = 11_2$. The complete codeword is therefore $11|100|1010|1111011001|0|n$. Notice how the length groups start with 1, which implies that each group is followed by a 1, except the last length group which is followed by a 0.

Decoding is simple. The decoder starts by reading the first $l$ bits. These are the length of the next group. More and more length groups are read, until a group is found that is followed by a 0. This indicates that the next group is $n$ itself.

With this background, the recursive definition of the $R_l$ prefixes is easy to read and understand. We use the notation $L = 1 + \lfloor \log_2 n \rfloor$ and denote by $B(n, l)$ the $l$-bit binary representation (beta code) of the integer $n$. Thus, $B(12, 5) = 01100$. For $l \geq 2$, the prefixes are defined by

$$R_l(n) = \begin{cases} B(n, l), & \text{for } 0 \leq n \leq 2^l - 1, \\ R_l(L)B(n, L), & \text{for } n \geq 2^l. \end{cases}$$

Those familiar with the Even–Rodeh code (Section 2.6) may already have realized that this code is identical to $R_3$. Furthermore, the Elias omega code (Section 2.4) is in-between $R_2$ and $R_3$ with two differences: (1) the omega code encodes the quantities $L_i - 1$ and (2) the 0 separator is placed to the right of $n$.

$R_2(985) = 11\ 100\ 1010\ 1111011001$             $R_2(31{,}925) = 11\ 100\ 1111\ 111110010110101$
$R_3(985) = \qquad 100\ 1010\ 1111011001$             $R_3(31{,}925) = \qquad 100\ 1111\ 111110010110101$
$R_4(985) = \qquad\quad 1010\ 1111011001$             $R_4(31{,}925) = \qquad\quad 1111\ 111110010110101$
$R_5(985) = \qquad\quad 01010\ 1111011001$             $R_5(31{,}925) = \qquad\quad 01111\ 111110010110101$
$R_6(985) = \qquad\quad 001010\ 1111011001$             $R_6(31{,}925) = \qquad\quad 001111\ 111110010110101$

The short table lists the $R_l(n)$ prefixes for $2 \le l \le 6$ and 985-bit and 31,925-bit integers. It is obvious that the shortest prefixes are obtained for parameter values $l = L = 1 + \lfloor \log_2 n \rfloor$. Larger values of $l$ result in slightly longer prefixes, while smaller values require more length groups and therefore result in much longer prefixes. Elementary algebraic manipulations indicate that the range of best lengths $L$ for a given parameter $l$ is given by $[2^s, 2^e - 1]$ where $s = 2^{l-1}$ and $e = 2^l - 1 = 2s - 1$ (this is the range of best lengths $L$, not best integers $n$). The following table lists these intervals for a few $l$ values and makes it clear that small parameters, such as 2 and 3, are sufficient for most practical applications of data compression. Parameter $l = 2$ is best for integers that are 2 to 7 bits long, while $l = 3$ is best for integers that are 8 to 127 bits long.

| $l$ | $s$ | $e$ | $2^s$ | $2^e - 1$ |
|---|---|---|---|---|
| 2 | 2 | 3 | 2 | 7 |
| 3 | 4 | 7 | 8 | 127 |
| 4 | 8 | 15 | 128 | 32,767 |
| 5 | 9 | 31 | 32,768 | 2,147,483,647 |

The second family of Stout codes is similar, but with different prefixes that are denoted by $S_l(n)$. For small $l$ values, this family offers some improvement over the $R_l$ codes. Specifically, it removes the slight redundancy that exists in the $R_l$ codes because a length group cannot be 0 (which is why a length group in the omega code encodes $L_i - 1$ and not $L_i$). The $S_l$ prefixes are similar to the $R_l$ prefixes with the difference that a length group for $L_i$ encodes $L_i - 1 - l$. Thus, $S_2(985)$, the prefix of a 985-bit integer $n$, starts with the 10-bit length group 1111011001 and prepends to it the length group for $10 - 1 - 2 = 7 = 111_2$. To this is prepended the length group for $3 - 1 - 2 = 0$ as the two bits 00. The result is the 15-bit prefix 00|111|1111011001, shorter than the 19 bits of $R_2(985)$. Another example is $S_3(985)$, which starts with the same 1111011001 and prepends to it the length group for $10 - 1 - 3 = 6 = 110_2$. The recursion stops at this point because 110 is an $l$-bit group. The result is the 13-bit codeword 110|1111011001, again shorter than the 17 bits of $R_3(985)$. The $S_l(n)$ prefixes are defined recursively by

$$S_l(n) = \begin{cases} B(n, l), & \text{for } 0 \le n \le 2^l - 1, \\ R_l(L - 1 - l)B(n, L), & \text{for } n \ge 2^l. \end{cases}$$

Table 2.20 lists some $S_2(n)$ and $S_3(n)$ prefixes and illustrates their regularity. Notice that the leftmost column lists the values of $L$, i.e., the lengths of the integers being encoded, and not the integers themselves. A length group retains its value until the group that follows it becomes all 1's, at which point the group increments by 1 and the group that follows is reset to $10\ldots0$. All the length groups, except perhaps the leftmost one, start with 1. This regular behavior is the result of the choice $L_i - 1 - l$.

The prefix $S_2(64)$, for example, starts with the 7-bit group $1000000 = 64$ and prepends to it $S_2(7 - 1 - 2) = S_2(4) = 00|100$. We stress again that the table lists only

| L | $S_2(n)$ | $S_3(n)$ |
|---|---|---|
| 1 | 01 | 001 |
| 2 | 10 | 010 |
| 3 | 11 | 011 |
| 4 | 00 100 | 100 |
| 5 | 00 101 | 101 |
| 6 | 00 110 | 110 |
| 7 | 00 111 | 111 |
| 8 | 01 1000 | 000 1000 |
| 15 | 01 1111 | 000 1111 |
| 16 | 10 10000 | 001 10000 |
| 32 | 11 100000 | 010 100000 |
| 64 | 00 100 1000000 | 011 1000000 |
| 128 | 00 101 10000000 | 100 10000000 |
| 256 | 00 110 100000000 | 101 100000000 |
| 512 | 00 111 1000000000 | 110 1000000000 |
| 1024 | 01 1000 10000000000 | 111 10000000000 |
| 2048 | 01 1001 100000000000 | 000 1000 100000000000 |

Table 2.20: $S_2(n)$ and $S_3(n)$ Codes.

the prefixes, not the complete codewords. Once this is understood, it is not hard to see that the second Stout code is a prefix code. Once a codeword has been assigned, it will not be the prefix of any other codeword. Thus, for example, the prefixes of all the codewords for 64-bit integers start with the prefix 00 100 of the 4-bit integers, but any codeword for a 4-bit integer has a 0 following the 00 100, whereas the codewords for the 64-bit integers have a 1 following 00 100.

# 2.12 Boldi–Vigna (ζ) Codes

The World Wide Web (WWW) was created in 1990 and currently, after only 17 years of existence, it completely pervades our lives. It is used by many people and is steadily growing and finding more applications. Formally, the WWW is a collection of inter-linked, hypertext documents (web pages) that can be sent over the Internet. Mathematically, the pages and hyperlinks may be viewed as nodes and edges in a vast directed graph (a webgraph) which itself has become an important object of study. Currently, the graph of the entire WWW consists of hundreds of millions of nodes and over a billion links. Experts often claim that these numbers grow exponentially with time. The various web search engines crawl over the WWW, collecting pages, resolving their hypertext links, and creating large webgraphs. Thus, a search engine operates on a vast data base and it is natural to want to compress this data.

One approach to compressing a webgraph is outlined in [Randall et al. 01]. Each URL is replaced by an integer. These integers are stored in a list that includes, for each integer, a pointer to an adjacency list of links. Thus, for example, URL 104 may be a web page that links to URLs 101, 132, and 174. These three integers become the adjacency list of 104, and this list is compressed by computing the differences of adjacent links and replacing the differences with appropriate variable-length codes. In our example, the differences are $101 - 104 = -3$, $132 - 101 = 31$, and $174 - 132 = 42$.

Experiments with large webgraphs indicate that the differences (also called gaps or deltas) computed for a large webgraph tend to be distributed according to a power law whose parameter in normally in the interval $[1.1, 1.3]$, so the problem of compressing a webgraph is reduced to the problem of finding a variable-length code that corresponds to such a distribution.

A power law distribution has the form

$$Z_\alpha[n] = \frac{P}{n^\alpha}$$

where $Z_\alpha[n]$ is the distribution (the number of occurrences of the value $n$), $\alpha$ is a parameter of the distribution, and $P$ is a constant of proportionality. Codes for power law distributions have already been mentioned. The length of the Elias gamma code (Equation (2.1)) is $2\lfloor \log_2 n \rfloor + 1$, which makes it a natural choice for a distribution of the form $1/(2n^2)$. This is a power law distribution with $\alpha = 2$ and $P = 1/2$. Similarly, the Elias delta code is suitable for a distribution of the form $1/[2n(\log_2(2n))^2]$. This is very close to a power law distribution with $\alpha = 1$. The nibble code and its variations (page 40) correspond to power law distributions of the form $1/n^{1+\frac{1}{k}}$, where the parameter is $1 + \frac{1}{k}$.

The remainder of this section describes the zeta ($\zeta$) code, also known as Boldi–Vigna code, introduced by Paolo Boldi and Sebastiano Vigna as a family of variable-length codes that are best choices for the compression of webgraphs. The original references are [Boldi and Vigna 04a] and [Boldi and Vigna 04b]. The latest reference is [Boldi and Vigna 05].

We start with Zipf's law, an empirical power law [Zipf 07] introduced by the linguist George K. Zipf. It states that the frequency of any word in a natural language is roughly inversely proportional to its position in the frequency table. Intuitively, Zipf's law states that the most frequent word in any language is twice as frequent as the second most-frequent word, which in turn is twice as frequent as the third word, and so on.

For a language having $N$ words, the Zipf power law distribution with a parameter $\alpha$ is given by

$$Z_\alpha[n] = \frac{1/n^\alpha}{\sum_{i=1}^{N} \frac{1}{i^\alpha}}$$

If the set is infinite, the denominator becomes the well-known Riemann zeta function

$$\zeta(\alpha) = \sum_{i=1}^{\infty} \frac{1}{i^\alpha}$$

which converges to a finite value for $\alpha > 1$. In this case, the distribution can be written in the form

$$Z_\alpha[n] = \frac{1}{\zeta(\alpha)n^\alpha}$$

The Boldi–Vigna zeta code starts with a positive integer $k$ that becomes the shrinking factor of the code. The set of all positive integers is partitioned into the intervals $[2^0, 2^k - 1]$, $[2^k, 2^{2k} - 1]$, $[2^{2k}, 2^{3k} - 1]$, and in general $[2^{hk}, 2^{(h+1)k} - 1]$. The length of such interval is $2^{(h+1)k} - 2^{hk}$.

Next, a minimal binary code is defined, which is closely related to the phased-in codes of Section 1.9. Given an interval $[0, z - 1]$ and an integer $x$ in this interval, we first compute $s = \lceil \log_2 z \rceil$. If $x < 2^s - z$, it is coded as the $x$th element of the interval, in $s - 1$ bits. Otherwise, it is coded as the $(x - z - 2^s)$th element of the interval in $s$ bits.

With this background, here is how the zeta code is constructed. Given a positive integer $n$ to be encoded, we employ $k$ to determine the interval where $n$ is located. Once this is known, the values of $h$ and $k$ are used in a simple way to construct the zeta code of $n$ in two parts, the value of $h + 1$ in unary (as $h$ zeros followed by a 1), followed by the minimal binary code of $n - 2^{hk}$ in the interval $[0, 2^{(h+1)k} - 2^{hk} - 1]$.

**Example.** Given $k = 3$ and $n = 16$, we first determine that $n$ is located in the interval $[2^3, 2^6 - 1]$, which corresponds to $h = 1$. Thus, $h + 1 = 2$ and the unary code of 2 is 01. The minimal binary code of $16 - 2^3 = 8$ is constructed in steps as follows. The length $z$ of the interval $[2^3, 2^6 - 1]$ is 56. This implies that $s = \lceil \log_2 56 \rceil = 6$. The value 8 to be encoded satisfies $8 = 2^6 - 56$, so it is encoded as $x - z - 2^s = 8 - 56 - 2^6 = 16$ in six bits, resulting in 010000. Thus, the $\zeta_3$ code of $n = 16$ is 01|010000.

Table 2.21 lists $\zeta$ codes for various shrinking factors $k$. For $k = 1$, the $\zeta_1$ code is identical to the $\gamma$ code. The nibble code of page 40 is also shown.

| $n$ | $\gamma = \zeta_1$ | $\zeta_2$ | $\zeta_3$ | $\zeta_4$ | $\delta$ | Nibble |
|---|---|---|---|---|---|---|
| 1 | 1 | 10 | 100 | 1000 | 1 | 1000 |
| 2 | 010 | 110 | 1010 | 10010 | 0100 | 1001 |
| 3 | 011 | 111 | 1011 | 10011 | 0101 | 1010 |
| 4 | 00100 | 01000 | 1100 | 10100 | 01100 | 1011 |
| 5 | 00101 | 01001 | 1101 | 10101 | 01101 | 1100 |
| 6 | 00110 | 01010 | 1110 | 10110 | 01110 | 1101 |
| 7 | 00111 | 01011 | 1111 | 10111 | 01111 | 1110 |
| 8 | 0001000 | 011000 | 0100000 | 11000 | 00100000 | 1111 |
| 9 | 0001001 | 011001 | 0100001 | 11001 | 00100001 | 00011000 |
| 10 | 0001010 | 011010 | 0100010 | 11010 | 00100010 | 00011001 |
| 11 | 0001011 | 011011 | 0100011 | 11011 | 00100011 | 00011010 |
| 12 | 0001100 | 011100 | 0100100 | 11100 | 00100100 | 00011011 |
| 13 | 0001101 | 011101 | 0100101 | 11101 | 00100101 | 00011100 |
| 14 | 0001110 | 011110 | 0100110 | 11110 | 00100110 | 00011101 |
| 15 | 0001111 | 011111 | 0100111 | 11111 | 00100111 | 00011110 |
| 16 | 000010000 | 00100000 | 01010000 | 010000111 | 001011001 | 00011111 |

Table 2.21: $\zeta$-Codes for $1 \leq k \leq 4$ Compared to $\gamma$, $\delta$ and Nibble Codes.

A nibble (more accurately, nybble) is the popular term for a 4-bit number (or half a byte). A nibble can therefore have 16 possible values, which makes it identical to a single hexadecimal digit

The length of the zeta code of $n$ with shrinking factor $k$ is $\lfloor 1 + (\log_2 n)/k \rfloor (k+1) + \tau(n)$ where

$$\tau(n) = \begin{cases} 0, & \text{if } (\log_2 n)/k - \lfloor (\log_2 n)/k \rfloor \in [0, 1/k), \\ 1, & \text{otherwise.} \end{cases}$$

Thus, this code is ideal for integers $n$ that are distributed as

$$\frac{1 + \tau(n)}{n^{1 + \frac{1}{k}}}.$$

This is very close to a power law distribution with a parameter $1 + \frac{1}{k}$. The developers show that the zeta code is complete, and they provide a detailed analysis of the expected lengths of the zeta code for various values of $k$. The final results are summarized in Table 2.22 where certain codes are recommended for various ranges of the parameter $\alpha$ of the distribution.

| $\alpha$ : | $< 1.06$ | $[1.06, 1.08]$ | $[1.08, 1.11]$ | $[1.11, 1.16]$ | $[1.16, 1.27]$ | $[1.27, 1.57]$ | $[1.57, 2]$ |
|---|---|---|---|---|---|---|---|
| Code: | $\delta$ | $\zeta_6$ | $\zeta_5$ | $\zeta_4$ | $\zeta_3$ | $\zeta_2$ | $\gamma = \zeta_1$ |

Table 2.22: Recommended Ranges for Codes.

# 2.13 Yamamoto's Recursive Code

In 2000, Hirosuke Yamamoto came up with a simple, ingenious way [Yamamoto 00] to improve the Elias omega code. As a short refresher on the omega code, we start with the relevant paragraph from Section 2.4.

> The omega code uses itself recursively to encode the prefix $M$, which is why it is sometimes referred to as a recursive Elias code. The main idea is to prepend the length of $n$ to $n$ as a group of bits that starts with a 1, then prepend the length of the length, as another group, to the result, and continue prepending lengths until the last length is 2 or 3 (and it fits in two bits). In order to distinguish between a length group and the last, rightmost group (that of $n$ itself), the latter is followed by a delimiter of 0, while each length group starts with a 1.

> The decoder of the omega code reads the first two bits (the leftmost length group), interprets its value as the length of the next group, and continues reading groups until a group is found that is followed by the 0 delimiter. That group is the binary representation of $n$.

> This scheme makes it easy to decode the omega code, but is somewhat wasteful, because the MSB of each length group is 1, and so the value of this bit is known in advance and it acts only as a separator.

> Yamamoto's idea was to select an $f$-bit delimiter $a$ (where $f$ is a small positive integer, typically 2 or 3) and use it as the delimiter instead of a single 0. In order to

obtain a UD code, none of the length groups should start with $a$. Thus, the length groups should be encoded with special variable-length codes that do not start with $a$. Once $a$ has been selected and its length $f$ is known, the first step is to prepare all the binary strings that do not start with $a$ and assign them in order the code string that indicates the finite number $B_{a,f}(i)$ of these of the positive integers. Now, if a length group has the value $n$, then $B_{a,f}(n)$ is placed in the codeword instead of the binary value of $n$.

We start with a simpler code that is denoted by $B_{a,f}(n)$. As an example, we select $f = 2$ and the 2-bit delimiter $a = 00$. The binary strings that do not start with 00 are the following: the two 1-bit strings 0 and 1; the three 2-bit strings 01, 10, and 11; the six 3-bit strings 010, 011, 100, 101, 110, and 111; the 12 4-bit strings 0100, 1001, through 1111 (i.e., the 16 4-bit strings minus the four that start with 00), and so on. These strings are assigned to the positive integers as listed in the second column of Table 2.23. The third column of this table lists the first 26 $B$ codes for $a = 100$.

| $n$ | $B_{a,f}(n)$ | | $\tilde{B}_{a,f}(n)$ | |
|---|---|---|---|---|
| | $a = 00$ | $a = 100$ | $a = 00$ | $a = 100$ |
| 1 | 0 | 0 | 1 | 0 |
| 2 | 1 | 1 | 01 | 00 |
| 3 | 01 | 00 | 10 | 01 |
| 4 | 10 | 01 | 11 | 11 |
| 5 | 11 | 10 | 010 | 000 |
| 6 | 010 | 11 | 011 | 001 |
| 7 | 011 | 000 | 100 | 010 |
| 8 | 100 | 001 | 101 | 011 |
| 9 | 101 | 010 | 110 | 101 |
| 10 | 110 | 011 | 111 | 110 |
| 11 | 111 | 101 | 0100 | 111 |
| 12 | 0100 | 110 | 0101 | 0000 |
| 13 | 0101 | 111 | 0110 | 0001 |
| 14 | 0110 | 0000 | 0111 | 0010 |
| 15 | 0111 | 0001 | 1000 | 0011 |
| 16 | 1000 | 0010 | 1001 | 0100 |
| 17 | 1001 | 0011 | 1010 | 0101 |
| 18 | 1010 | 0100 | 1011 | 0110 |
| 19 | 1011 | 0101 | 1100 | 0111 |
| 20 | 1100 | 0110 | 1101 | 1010 |
| 21 | 1101 | 0111 | 1110 | 1011 |
| 22 | 1110 | 1010 | 1111 | 1100 |
| 23 | 1111 | 1011 | 01000 | 1101 |
| 24 | 01000 | 1100 | 01001 | 1110 |
| 25 | 01001 | 1101 | 01010 | 1111 |
| 26 | 01010 | 1110 | 01011 | 00000 |

Table 2.23: Some $B_{a,f}(n)$ and $\tilde{B}_{a,f}(n)$ Yamamoto Codes.

One more step is needed. Even though none of the $B_{a,f}(n)$ codes starts with $a$, it may happen that a short $B_{a,f}(n)$ code (a code whose length is less than $f$) followed by another $B_{a,f}(n)$ code will accidentally contain the pattern $a$. Thus, for example, the table shows that $B_{100,3}(5) = 10$ followed by $B_{100,3}(7) = 000$ becomes the string $10|000$ and may be interpreted by the decoder as $100|00\ldots$. Thus, the $B_{100,3}(n)$ codes have to be made UD by eliminating some of the short codes, those that are shorter than $f$ bits and have the values 1 or 10. Similarly, the single short $B_{00,2}(1) = 0$ code should be discarded. The resulting codes are designated $\tilde{B}_{a,f}(n)$ and are listed in the last two columns of the table.

If the integers to be encoded are known not to exceed a few hundred or a few thousand, then both encoder and decoder can have built-in tables of the $\tilde{B}_{a,f}(n)$ codes for all the relevant integers. In general, these codes have to be computed "on the fly" and the developer provides the following expression for them (the notation $K^{[j]}$ means the integer $K$ expressed in $j$ bits, where some of the leftmost bits may be zeros).

$$\tilde{B}_{a,f}(n) = \begin{cases} [n - M(j,f) + L(j,f)]^{[j]}, \\ \qquad \text{if } M(j,f) - L(j,f) \le n < M(j,f) - L(j,f) + N(j,f,a), \\ [n - M(j,f-1) + L(j+1,f)]^{[j]}, \\ \qquad \text{if } M(j,f) - L(j,f) + N(j,f,a) \le n < M(j+1,f) - L(j+1,f), \end{cases}$$

where $M(j,f) = 2^j - 2^{(j-f)_+}$, $(t)_+ = \max(t,0)$, $L(j,f) = (f-1) - (f-j)_+$, and $N(j,f,a) = \lfloor 2^{j-f}a \rfloor$. Given a positive integer $n$, the first step in encoding it is to determine the value of $j$ by examining the inequalities. Once $j$ is known, functions $M$ and $L$ can be computed.

Armed with the $\tilde{B}_{a,f}(n)$ codes, the recursive Yamamoto code $C_{a,f}(n)$ is easy to describe. We select an $f$-bit delimiter $a$. Given a positive integer $n$, we start with the group $\tilde{B}_{a,f}(n)$. Assuming that this group is $n_1$ bits long, we prepend to it the length group $\tilde{B}_{a,f}(n_1 - 1)$. If this group is $n_2$ bits long, we prepend to it the length group $\tilde{B}_{a,f}(n_2 - 1)$. This is repeated recursively until the last length group is $\tilde{B}_{a,f}(1)$. (This code is always a single bit because it depends only on $a$, and the decoder knows this bit because it knows $a$. Thus, in principle, the last length group can be omitted, thereby shortening the codeword by one bit.) The codeword is completed by appending the delimiter $a$ to the right end. Table 2.24 lists some codewords for $a = 00$ and for $a = 100$.

The developer, Hirosuke Yamamoto, also proves that the length of a codeword $C_{a,f}(n)$ is always less than or equal to $\log_2^* n + F(f)w_f(n) + c_f + \delta(n)$. More importantly, for infinitely many integers, the length is also less than or equal to $\log_2^* n + (1 - F(f))w_f(n) + c_f + 2\delta(n)$, where $F(f) = -\log_2(1 - 2^{-f})$, $c_f = 5(f-2)_+ + f + 5F(f)$,

$$\delta(n) \le \frac{\log_2 e}{n}\left[1 + \frac{(w(n)-1)(\log_2 e)^{w(n)-1}}{\log_2 n}\right] \le \frac{4.7}{n},$$

and the log-star function $\log_2^* n$ is the finite sum

$$\log_2 n + \log_2 \log_2 n + \cdots + \underbrace{\log_2 \log_2 \ldots \log_2 n}_{w(n)}$$

where $w(n)$ is the largest integer for which the compound logarithm is still nonnegative.

| $n$ | $a = 00$ | $a = 100$ |
|---|---|---|
| 1 | 1 00 | 0 100 |
| 2 | 1 01 00 | 0 00 100 |
| 3 | 1 10 00 | 0 01 100 |
| 4 | 1 11 00 | 0 11 100 |
| 5 | 1 01 010 00 | 0 00 000 100 |
| 6 | 1 01 011 00 | 0 00 001 100 |
| 7 | 1 01 100 00 | 0 00 010 100 |
| 8 | 1 01 101 00 | 0 00 011 100 |
| 9 | 1 01 110 00 | 0 00 101 100 |
| 10 | 1 01 111 00 | 0 00 110 100 |
| 11 | 1 10 0100 00 | 0 00 111 100 |
| 12 | 1 10 0101 00 | 0 01 0000 100 |
| 13 | 1 10 0110 00 | 0 01 0001 100 |
| 14 | 1 10 0111 00 | 0 01 0010 100 |
| 15 | 1 10 1000 00 | 0 01 0011 100 |
| 16 | 1 10 1001 00 | 0 01 0100 100 |
| 17 | 1 10 1010 00 | 0 01 0101 100 |
| 18 | 1 10 1011 00 | 0 01 0110 100 |
| 19 | 1 10 1100 00 | 0 01 0111 100 |
| 20 | 1 10 1101 00 | 0 01 1010 100 |
| 21 | 1 10 1110 00 | 0 01 1011 100 |
| 22 | 1 10 1111 00 | 0 01 1100 100 |
| 23 | 1 11 01000 00 | 0 01 1101 100 |
| 24 | 1 11 01001 00 | 0 01 1110 100 |
| 25 | 1 11 01010 00 | 0 01 1111 100 |
| 26 | 1 11 01011 00 | 0 11 00000 100 |

Table 2.24: Yamamoto Recursive Codewords.

The important point (at least from a theoretical point of view) is that this recursive code is shorter than $\log_2^* n$ for infinitely many integers. This code can also be extended to other number bases and is not limited to base-2 numbers.

# 2.14 VLCs and Search Trees

There is an interesting association between certain unbounded searches and prefix codes. A search is a classic problem in computer science. Given a data structure, such as an array, a list, a tree, or a graph, where each node contains an item, the problem is to search the structure and find a given item in the smallest possible number of steps. The following are practical examples of computer searches.

1. A two-player game. Player $A$ chooses a positive integer $n$ and player $B$ has to guess $n$ by asking only the following type of question: Is the integer $i$ less than $n$? Clearly, the number of questions needed to find $n$ depends on $n$ and on the strategy (algorithm) used by $B$.

2. Given a function $y = f(x)$, find all the values of $x$ for which the function equals a given value $Y$ (typically zero).

3. An online business stocks many items. Customers should be able to search for an item by its name, price range, or manufacturer. The items are the nodes of a large structure (normally a tree), and there must be a fast search algorithm that can find any item among many thousands of items in a few seconds.

4. An Internet search engine faces a similar problem. The first part of such an engine is a crawler that locates a vast number of Web sites, inputs them into a data base, and indexes every term found. The second part is a search algorithm that should be able to locate any term (out of many millions of terms) in just a few seconds. (There is also the problem of ranking the results.) Internet users search for terms (such as "variable-length codes") and naturally prefer search engines that provide results in just a few seconds.

Searching a data base, even a very large one, is considered a bounded search, because the search space is finite. Bounded searches have been a popular field of research for many years, and many efficient search methods are known and have been analyzed in detail. In contrast, searching for a zero of a function is an example of unbounded search, because the domain of the function is normally infinite. In their work [Bentley and Yao 76], Jon Bentley and Andrew Yao propose several algorithms for searching a linearly ordered unbounded table and show that each algorithm is associated with a binary string that can be considered a codeword in a prefix code. This creates an interesting and unexpected relation between searching and variable-length codes, two hitherto unrelated areas of scientific endeavor. The authors restrict themselves to algorithms that search for an integer $n$ in an ordered table $F(i)$ using only elementary tests of the form $F(i) < n$.

The simplest search method, for both bounded and unbounded searches, is linear search. Given an ordered array of integers $F(i)$ and an integer $n$, a linear search performs the tests $F(1) < n$, $F(2) < n,\ldots$ until a match is found. The answers to the tests are No, No,$\ldots$, Yes, which can be expressed as the bitstring $11\ldots10$. Thus, a linear search (referred to by its developers as algorithm $B_0$) corresponds to the unary code.

The next search algorithm (designated $B_1$) is an unbounded variation of the well-known binary search. The first step of this version is a linear search that determines an interval. This step starts with an interval $[F(a), F(b)]$ and performs the test $F(b) < n$. If the answer is No, then $n$ is in this interval, and the algorithm executes its second step, where $n$ is located in this interval with a bounded binary search. If the answer is Yes, then the algorithm computes another interval that starts at $F(b+1)$ and is twice as wide as the previous interval. Specifically, the algorithm computes $F(2^i - 1)$ for $i = 1$, $2,\ldots$ and each interval is $[F(2^{i-1}), F(2^i - 1)]$.

Figure 2.25 (compare with Figures 1.12 and 1.13) shows how these intervals can be represented as a binary search tree where the root and each of its right descendants correspond to an index $2^i - 1$. The left son of each descendant is the root of a subtree that contains the remaining indexes less than or equal to $2^i - 1$. Thus, the left son of the root is index 1. The left son of node 3 is the root of a subtree containing the remaining indexes that are less than or equal to 3, namely 2 and 3. The left son of 7 is the root of a subtree with 4, 5, 6, and 7, and so on.

The figure also illustrates how $n = 12$ is searched for and located by the $B_1$ method. The algorithm compares 12 to 1, 3, 7, and 15 (a linear search), with results No, No, No,

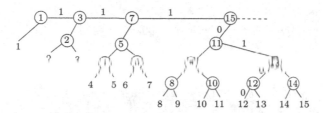

Figure 2.25: Gamma Code and Binary Search.

and Yes, or 1110. The next step performs a binary search to locate 12 in the interval $[8, 15]$ with results 100. These results are listed as labels on the appropriate edges. The final bitstring associated with 12 by this method is therefore $1110|100$. This result is the Elias gamma code of 12, as produced by the alternative construction method of page 74 (this construction produces codes different from those of Table 2.8, but the code lengths are the same).

If $m$ is the largest integer such that $2^{m-1} \le n < 2^m$, then the first step (a linear search to determine $m$) requires $m = \lfloor \log_2 n \rfloor + 1$ tests. The second step requires $\log_2 2^{m-1} = m - 1 = 2\lfloor \log_2 n \rfloor$ tests. The total number of tests is therefore $2\lfloor \log_2 n \rfloor + 1$, which is the length of the gamma code (Equation (2.1)).

The third search method proposed by Bentley and Yao (their algorithm $B_2$) is a double binary search. The second step is the same, but the first step is modified to determine $m$ by constructing intervals that become longer and longer. The intervals computed by method $B_1$ start at $2^i - 1$ for $i = 1, 2, 3, \ldots$, or 1, 3, 7, 15, 31, $\ldots$. Method $B_2$ constructs intervals that start with numbers of the form $2^{2^{i-1}} - 1$ for $i = 1$, 2, 3, $\ldots$. These numbers are 1, 7, 127, 32,767, $2^{31} - 1$, and so on. They become the roots of subtrees as shown in Figure 2.26. There are fewer search trees, which speeds up step 1, but they grow very fast. The developers show that the first step requires $1 + 2\lfloor \log_2 n \rfloor$ tests (compared to the $m = \lfloor \log_2 n \rfloor + 1$ tests required by $B_1$) and the second step requires $\lfloor \log_2 n \rfloor$ tests (same as $B_1$), for a total of

$$\lfloor \log_2 n \rfloor + 2\lfloor \log_2(\lfloor \log_2 n \rfloor + 1) \rfloor + 1 = 1 + \lfloor \log_2 n \rfloor + 2\lfloor \log_2 \log_2(2n) \rfloor.$$

The number of tests is identical to the length of the Elias delta code, Equation (2.2). Thus, the variable-length code associated with this search tree is a variant of the delta code. The codewords are different from those listed in Table 2.9, but the code is equivalent because of the identical lengths.

Bentley and Yao go on to propose unbounded search algorithms $B_k$ ($k$-nested binary search) and $U$ (ultimate). The latter is shown by [Ahlswede et al. 97] to be associated with a modified Elias omega code. The authors of this paper also construct the binary search trees for the Stout codes (Section 2.11).

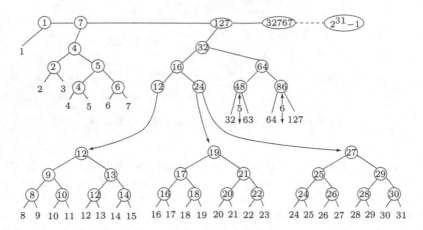

Figure 2.26: Delta Code and Binary Search.

# 2.15 Taboo Codes

The taboo approach to variable-length codes, as well as the use of the term "taboo," are the brainchilds of Steven Pigeon. The two types of taboo codes are described in [Pigeon 01a,b] where it is shown that they are universal (more accurately, they can be made as close to universal as desired by the choice of a parameter). The principle of the taboo codes is to select a positive integer parameter $n$ and reserve a pattern of $n$ bits to indicate the end of a code. This pattern should not appear in the code itself, which is the reason for the term taboo. Thus, the taboo codes can be considered suffix codes.

The first type of taboo code is block-based and its length is a multiple of $n$. The block-based taboo code of an integer is a string of $n$-bit blocks, where $n$ is a user-selected parameter and the last block is a taboo bit pattern that cannot appear in any of the other blocks. An $n$-bit block can have $2^n$ values, so if one value is reserved for the taboo pattern, each of the remaining code blocks can have one of the remaining $2^n - 1$ bit patterns. In the second type of taboo codes, the total length of the code is not restricted to a multiple of $n$. This type is called unconstrained and is shown to be related to the $n$-step Fibonacci numbers.

We use the notation $\langle n \rangle : t$ to denote a string of $n$ bits concatenated with the taboo string $t$. Table 2.27 lists the lengths, number of codes, and code ranges as the codes get longer when more blocks are added. Each row in this table summarizes the properties of a *range* of codes. The number of codes in the $k$th range is $(2^n - 1)^k$, and the total number of codes $g_n(k)$ in the first $k$ ranges is obtained as the sum of a geometric progression

$$g_n(k) = \sum_{i=1}^{k}(2^n - 1)^i = \frac{[(2^n - 1)^k - 1](2^n - 1)}{2^n - 2}.$$

The case $n = 1$ is special and uninteresting. A 1-bit block can either be a 0 or a 1. If we reserve 0 as the taboo pattern, then all the other blocks of codes cannot contain any zeros and must be all 1's. The result is the infinite set of codewords **10, 110, 1110,...**,

| Codes | Length | # of values | Range |
|-------|--------|-------------|-------|
| $\langle n{:}t \rangle$ | $2n$ | $2^n - 1$ | $0$ to $(2^n - 1) - 1$ |
| $\langle 2n{:}t \rangle$ | $3n$ | $(2^n - 1)^2$ | $(2^n - 1)$ to $(2^n - 1) + (2^n - 1)^2 - 1$ |
| $\vdots$ | | | $\vdots$ |
| $\langle kn{:}t \rangle$ | $(k+1)n$ | $(2^n - 1)^k$ | $\sum_{i=1}^{k-1}(2^n - 1)^i$ to $-1 + \sum_{i=1}^{k}(2^n - 1)^i$ |
| $\vdots$ | | | $\vdots$ |

Table 2.27: Structure of Block-Based Taboo Codes.

which is the unary code and therefore not universal (more accurately, the unary code is $\infty$-universal). The first interesting case is $n = 2$. Table 2.28 lists some codes for this case. Notice that the pattern of all zeros is reserved for the taboo, but any other $n$-bit pattern would do as well.

| $m$ | Code | $m$ | Code | $m$ | Code | $m$ | Code |
|-----|------|-----|------|-----|------|-----|------|
| 0 | 01 **00** | 4 | 01 10 **00** | 8 | 10 11 **00** | 12 | 01 01 01 **00** |
| 1 | 10 **00** | 5 | 01 11 **00** | 9 | 11 01 **00** | 13 | 01 01 10 **00** |
| 2 | 11 **00** | 6 | 10 01 **00** | 10 | 11 10 **00** | 14 | 01 01 11 **00** |
| 3 | 01 01 **00** | 7 | 10 10 **00** | 11 | 11 11 **00** | $\cdots$ | |

Table 2.28: Some Block-Based Taboo Codes for $n = 2$.

In order to use this code in practice, we need simple, fast algorithms to encode and decode it. Given an integer $m$, we first determine the number of $n$-bit blocks in the code of $m$. Assuming that $m$ appears in the $k$th range of Table 2.27, we can write the basic relation $m \le g_n(k) - 1$. This can be written explicitly as

$$m \le \frac{[(2^n - 1)^k - 1](2^n - 1)}{2^n - 2} - 1$$

or

$$k \ge \frac{\log_2\left[m + 2 - \frac{m+1}{2^n - 1}\right]}{\log_2(2^n - 1)}.$$

The requirement that $m$ be less than $g_n(k)$ yields the value of $k$ (which depends on both $n$ and $m$) as

$$k_n(m) = \left\lceil \frac{\log_2\left[m + 2 - \frac{m+1}{2^n - 1}\right]}{\log_2(2^n - 1)} \right\rceil.$$

Thus, the code of $m$ consists of $k_n(m)$ blocks of $n$ bits each plus the $n$-bit taboo block, for a total of $[k_n(m) + 1]n$ bits. This code can be considered the value of the integer $c \overset{\text{def}}{=} m - g_n(k_n(m) - 1)$ represented in base $(2^n - 1)$, where each "digit" is $n$ bits long.

**Encoding**: Armed with the values of $n$, $m$, and $c$, we can write the individual values of the $n$-bit blocks $b_i$ that constitute the code of $m$ as

$$b_i = \left[\lceil c(2^n - 1)^i \rceil \bmod (2^n - 1)\right] + 1, \quad i = 0, 1, \ldots, k_n(m) - 1. \tag{2.3}$$

The taboo pattern is the $(k_n(m) + 1)$th block, block $b_{k_n(m)}$. Equation (2.3) is the basic relation for encoding integer $m$.

**Decoding**: The decoder knows the value of $n$ and the taboo pattern. It reads $n$-bit blocks $b_i$ until it finds the taboo block. The number of blocks read is $k$ (rather $k_n(m)$) and once $k$ is known, decoding is done by

$$m = g_n(k - 1) + \sum_{i=0}^{k-1} (b_i - 1)(2^n - 1)^i. \tag{2.4}$$

Equations (2.3) and (2.4) lend themselves to fast, efficient implementation. In particular, decoding is easy. The decoder can read $n$-bit blocks $b_i$ and add terms to the partial sum of Equation (2.4) until it reaches the taboo block, at which point it only has to compute and add $g_n(k_n(m) - 1)$ to the sum in order to obtain $m$.

The block-based taboo code is longer than the binary representation (beta code) of the integers by $n + k\lceil n - \log_2(2^n - 1)\rceil$ bits. For all values of $n$ except the smallest ones, $2^n - 1 \approx 2^n$, so this difference of lengths (the "waste" of the code) equals approximately $n$. Even for $n = 4$, the waste is only $n + 0.093k$. Thus, for large integers $m$, the waste is very small compared to $m$.

The developer of these codes, Steven Pigeon, proves that the block-based taboo code can be made as close to universal as desired.

> Let us overthrow the totems, break the taboos. Or better, let us consider them cancelled. Coldly, let us be intelligent.
>
> —Pierre Trudeau

Now, for the unconstrained taboo codes. The idea is to have a bitstring of length $l$, the actual code, followed by a taboo pattern of $n$ bits. The length $l$ of the code is independent of $n$ and can even be zero. Thus, the total length of such a code is $l + n$ for $l \geq 0$. The taboo pattern should not appear inside the $l$ code bits and there should be a separator bit at the end of the $l$-bit code part, to indicate to the decoder the start of the taboo. In the block-based taboo codes, any bit pattern could serve as the taboo, but for the unconstrained codes we have to choose a bit pattern that will prevent ambiguous decoding. The two ideal patterns for the taboo are therefore all zeros or all 1's, because a string of all zeros can be separated from the preceding $l$ bits by a single 1, and similarly for a string of all 1's. Other strings require longer separators. Thus, the use of all zeros or 1's as the taboo string leads to the maximum number of $l$-bit valid code strings and we will therefore assume that the taboo pattern consists of $n$ consecutive zeros.

As a result, an unconstrained taboo code starts with $l$ bits (where $l \geq 0$) that do not include $n$ consecutive zeros and that end with a separator bit of 1 (except when $l = 0$, where the code part consists of no bits and there is no need for a separator). The case $n = 1$ is special. In this case, the taboo is a single zero, the $l$ bits preceding it cannot include any zeros, so the unconstrained code reduces to the unary code of $l$ 1's followed by a zero.

In order to figure out how to encode and decode these codes, we first have to ⎯⎯⎯⎯⎯⎯⎯⎯⎯⎯⎯⎯⎯⎯⎯⎯⎯⎯⎯⎯⎯⎯⎯⎯⎯⎯⎯⎯⎯⎯⎯⎯⎯⎯ any $n$ consecutive zeros. We distinguish three cases.

1. The case $l = 0$ is trivial. The number of bit patterns of length zero is one, the pattern of no bits.

2. The case $0 < l < n$ is simple. The string of $l$ bits is too short to have any $n$ consecutive zeros. The last bit must be a 1, so the remaining $l - 1$ bits can have $2^{l-1}$ values.

3. When $l \geq n$, we determine the number of valid $l$-bit strings (i.e., $l$-bit strings that do not contain any substrings of $n$ zeros) recursively. We denote the number of valid strings by $\left\langle\!\left\langle {l \atop n} \right\rangle\!\right\rangle$ and consider the following cases. When a valid $l$-bit string starts with a 1 followed by $l-1$ bits, the number of valid strings is $\left\langle\!\left\langle {l-1 \atop n} \right\rangle\!\right\rangle$. When such a string start with a 01, the number of valid strings is $\left\langle\!\left\langle {l-2 \atop n} \right\rangle\!\right\rangle$. We continue in this way until we reach strings that start with $\underbrace{00\ldots0}_{n-1}1$, where the number of valid strings is $\left\langle\!\left\langle {l-n \atop n} \right\rangle\!\right\rangle$.

Thus, the number of valid $l$-bit strings in this case is the sum shown in the third case of Equation (2.5).

$$\left\langle\!\left\langle {l \atop n} \right\rangle\!\right\rangle = \begin{cases} 1 & \text{if } l = 0, \\ 2^{l-1} & \text{if } l < n, \\ \sum_{i=1}^{n} \left\langle\!\left\langle {l-i \atop n} \right\rangle\!\right\rangle & \text{otherwise.} \end{cases} \tag{2.5}$$

Table 2.29 lists some values of $\left\langle\!\left\langle {l \atop n} \right\rangle\!\right\rangle$ and it is obvious that the second column (the values of $\left\langle\!\left\langle {l \atop 2} \right\rangle\!\right\rangle$) consists of the well-known Fibonacci numbers. Thus, $\left\langle\!\left\langle {l \atop 2} \right\rangle\!\right\rangle = F_{l+1}$. A deeper look at the table shows that $\left\langle\!\left\langle {l \atop 3} \right\rangle\!\right\rangle = F_{l+1}^{(3)}$, where $F_l^{(3)} = F_{l-1}^{(3)} + F_{l-2}^{(3)} + F_{l-3}^{(3)}$. This column consists of the $n$-step Fibonacci numbers of order 3 (tribonaccis), a sequence that starts with $F_1^{(3)} = 1$, $F_2^{(3)} = 1$, and $F_3^{(3)} = 2$. Similarly, the 4th column consists of tetrabonacci numbers, and the remaining columns feature the pentanacci, hexanacci, heptanacci, and other $n$-step "polynacci" numbers.

The $n$-step Fibonacci numbers of order $k$ are defined recursively by

$$F_l^{(k)} = \sum_{i=1}^{k} F(k)_{l-i}, \tag{2.6}$$

with initial conditions $F(k)_1 = 1$ and $F(k)_i = 2^{i-2}$ for $i = 2, 3, \ldots, k$. (As an aside, the limit $\lim_{k \to \infty} F^{(k)}$ is a sequence that starts with an infinite number of 1's, followed by the powers of 2. The interested reader should also look for anti-Fibonacci numbers, generalized Fibonacci numbers, and other sequences and numbers related to Fibonacci.) We therefore conclude that $\left\langle\!\left\langle {l \atop n} \right\rangle\!\right\rangle = F_{l+1}^{(n)}$, a relation that is exploited by the developer of these codes to prove that the unconstrained taboo codes are universal (more accurately, can be made as close to universal as desired by increasing $n$).

Table 2.30 lists the organization, lengths, number of codes, and code ranges of some unconstrained taboo codes, and Table 2.31 lists some of these codes for $n = 3$ (the taboo string is shown in boldface).

# 2. Advanced Codes

| | $\left\langle\!\!\left\langle {l\atop 1}\right\rangle\!\!\right\rangle$ | $\left\langle\!\!\left\langle {l\atop 2}\right\rangle\!\!\right\rangle$ | $\left\langle\!\!\left\langle {l\atop 3}\right\rangle\!\!\right\rangle$ | $\left\langle\!\!\left\langle {l\atop 4}\right\rangle\!\!\right\rangle$ | $\left\langle\!\!\left\langle {l\atop 5}\right\rangle\!\!\right\rangle$ | $\left\langle\!\!\left\langle {l\atop 6}\right\rangle\!\!\right\rangle$ | $\left\langle\!\!\left\langle {l\atop 7}\right\rangle\!\!\right\rangle$ | $\left\langle\!\!\left\langle {l\atop 8}\right\rangle\!\!\right\rangle$ |
|---|---|---|---|---|---|---|---|---|
| $\left\langle\!\!\left\langle {0\atop n}\right\rangle\!\!\right\rangle$ | 1 | 1 | 1 | 1 | 1 | 1 | 1 | 1 |
| $\left\langle\!\!\left\langle {1\atop n}\right\rangle\!\!\right\rangle$ | 1 | 1 | 1 | 1 | 1 | 1 | 1 | 1 |
| $\left\langle\!\!\left\langle {2\atop n}\right\rangle\!\!\right\rangle$ | 1 | 2 | 2 | 2 | 2 | 2 | 2 | 2 |
| $\left\langle\!\!\left\langle {3\atop n}\right\rangle\!\!\right\rangle$ | 1 | 3 | 4 | 4 | 4 | 4 | 4 | 4 |
| $\left\langle\!\!\left\langle {4\atop n}\right\rangle\!\!\right\rangle$ | 1 | 5 | 7 | 8 | 8 | 8 | 8 | 8 |
| $\left\langle\!\!\left\langle {5\atop n}\right\rangle\!\!\right\rangle$ | 1 | 8 | 13 | 15 | 16 | 16 | 16 | 16 |
| $\left\langle\!\!\left\langle {6\atop n}\right\rangle\!\!\right\rangle$ | 1 | 13 | 24 | 29 | 31 | 32 | 32 | 32 |
| $\left\langle\!\!\left\langle {7\atop n}\right\rangle\!\!\right\rangle$ | 1 | 21 | 44 | 56 | 61 | 63 | 64 | 64 |
| $\left\langle\!\!\left\langle {8\atop n}\right\rangle\!\!\right\rangle$ | 1 | 34 | 81 | 108 | 120 | 125 | 127 | 128 |
| $\left\langle\!\!\left\langle {9\atop n}\right\rangle\!\!\right\rangle$ | 1 | 55 | 149 | 208 | 236 | 248 | 253 | 255 |
| $\left\langle\!\!\left\langle {10\atop n}\right\rangle\!\!\right\rangle$ | 1 | 89 | 274 | 401 | 464 | 492 | 504 | 509 |

Table 2.29: The First Few Values of $\left\langle\!\!\left\langle {l\atop n}\right\rangle\!\!\right\rangle$.

| Codes | Length | # of values | Range |
|---|---|---|---|
| $t$ | $n$ | $\left\langle\!\!\left\langle {0\atop n}\right\rangle\!\!\right\rangle$ | 0 to $\left\langle\!\!\left\langle {0\atop n}\right\rangle\!\!\right\rangle - 1$ |
| $\langle 1\rangle{:}t$ | $1+n$ | $\left\langle\!\!\left\langle {1\atop n}\right\rangle\!\!\right\rangle$ | $\left\langle\!\!\left\langle {0\atop n}\right\rangle\!\!\right\rangle$ to $\left\langle\!\!\left\langle {0\atop n}\right\rangle\!\!\right\rangle + \left\langle\!\!\left\langle {1\atop n}\right\rangle\!\!\right\rangle - 1$ |
| $\vdots$ | | | $\vdots$ |
| $\langle l\rangle{:}t$ | $l+n$ | $\left\langle\!\!\left\langle {l\atop n}\right\rangle\!\!\right\rangle$ | $\sum_{i=0}^{l-1}\left\langle\!\!\left\langle {i\atop n}\right\rangle\!\!\right\rangle$ to $-1+\sum_{i=0}^{l}\left\langle\!\!\left\langle {i\atop n}\right\rangle\!\!\right\rangle$ |
| $\vdots$ | | | $\vdots$ |

Table 2.30: Organization and Features of Unconstrained Taboo Codes.

| $m$ | Code | $m$ | Code | $m$ | Code | $m$ | Code |
|---|---|---|---|---|---|---|---|
| 0 | **000** | 7 | 111**000** | 14 | 1111**000** | 21 | 10011**000** |
| 1 | 1**000** | 8 | 0011**000** | 15 | 00101**000** | 22 | 10101**000** |
| 2 | 01**000** | 9 | 0101**000** | 16 | 00111**000** | 23 | 10111**000** |
| 3 | 11**000** | 10 | 0111**000** | 17 | 01001**000** | 24 | 11001**000** |
| 4 | 001**000** | 11 | 1001**000** | 18 | 01011**000** | 25 | 11011**000** |
| 5 | 011**000** | 12 | 1011**000** | 19 | 01101**000** | 26 | 11101**000** |
| 6 | 101**000** | 13 | 1101**000** | 20 | 01111**000** | 27 | ... |

Table 2.31: Unconstrained Taboo Codes for $n = 3$.

> And I promise you, right here and now, no subject will ever be taboo... except, of course, the subject that was just under discussion.
> —Quentin Tarantino, *Kill Bill*, Vol. 1

The main reference [Pigeon 01b] describes the steps for encoding and decoding these codes.

# 2.16 Wang's Flag Code

Similar to the taboo code, the flag code of Muzhong Wang [Wang 88] is based on a flag of $f$ zeros appended to the codewords. The name *suffix* code is perhaps more appropriate, because the flag must be the suffix of a codeword and cannot appear anywhere inside it. However, this type of suffix code is not the opposite of a prefix code.

The principle is to scan the positive integer $n$ that is being encoded and append a single 1 to each sequence of $f - 1$ zeros found in it. This effectively removes any occurrences of substrings of $f$ consecutive zeros. Before this is done, $n$ is reversed, so that its LSB becomes a 1. This guarantees that the flag will be preceded by a 1 and will therefore be easily recognized by the decoder.

We use the notation $0^f$ for a string of $f$ zeros, select a value $f \geq 2$, and look at two examples. Given $n = 66 = 1000010_2$ and $f = 3$, we reverse $n$ to obtain 0100001 and scan it from left to right, appending a 1 to each string of two consecutive zeros. The result is 010010011 to which is appended the flag. Thus, 010010011|000. Given $n = 288 = 100100000_2$, we reverse it to 000001001, scan it and append 001001010011, and append the flag 001001010011|000.

It is obvious that such codewords are easy to decode. The decoder knows the value of $f$ and looks for a string of $f - 1$ zeros. If such a string is followed by a 0, then it is the flag. The flag is removed and the result is reversed. If such a string is followed by a 1, the 1 is removed and the scan continues.

This code, as originally proposed by Wang, is slow to implement in software because reversing an $m$-bit string requires $\lfloor m/2 \rfloor$ steps where a pair of bits is swapped in each step. A possible improvement is to move the MSB (which is always a 1) to the right end, where it acts as an intercalary bit to separate the code bits from the flag.

Reference [Wang 88] shows that this code is universal, and for large values of $n$ its asymptotic efficiency, defined by Wang as the limit

$$\Gamma = \lim_{n \to \infty} \sup \frac{\log_2(n+1)}{L(n)}, \tag{2.7}$$

(where $L(n)$ is the length of the Wang code of $n$) approaches 1.

The codeword length $L(n)$ increases with $n$, but not monotonically because it depends on the number of consecutive zeros in $n$ and on the relation of this number to $f$. Recall that the binary representation of a power of 2 is a single 1 followed by several consecutive zeros. Thus, integers of the form $2^k$ or slightly bigger tend to have many consecutive zeros. As a result, the Wang codewords of $2^k$ and of its immediate successors tend to be longer than the codewords of its immediate predecessors because of the many intercalary 1's. On the other hand, the binary representation of the integer $2^k - 1$ consists of $k$ consecutive 1's, so such numbers and their immediate predecessors tend to have few consecutive zeros, which is why their Wang codewords tend to be shorter. We therefore conclude that the lengths of Wang codewords feature jumps at $n$ values that are powers of 2.

It is easy to figure out the length of the Wang code for two types of numbers. The integer $n = 2^k - 1 = 1^k$ has no zeros in its binary representation, so its Wang code consists of the original $k$ bits plus the $f$-bit flag. Its successor $n + 1 = 2^k = 10^k$, on the other hand, has $k$ consecutive zeros, so $\lfloor k/f \rfloor$ intercalary bits have to be inserted. The length of the codeword is therefore the original $k+1$ bits, plus the extra $\lfloor k/f \rfloor$ bits, plus the flag, for a total of $k + 1 + \lfloor k/f \rfloor + f$. Thus, when $n$ passes through a value $2^k$, the codeword length increases by $\lfloor k/f \rfloor + 1$ bits.

Reference [Yamamoto and Ochi 91] shows that in the special case where each bit $b_j$ of $n$ (except the MSB $b_0$, which is effectively unused by this code) is selected independently with probability $P(b_j = 0) = P(b_j = 1) = 1/2$ for $j = 1, 2, \ldots, M$, the average codeword length $\overline{L(n)}$ depends on $M$ and $f$ in a simple way (compare with Equation (2.9))

$$\overline{L(n)} = \begin{cases} M + 1 + f & \text{if } M \leq f - 2, \\ M + 1 + \sum_{j=1}^{M-f+2} l_j + f, & \text{if } M \geq f - 1, \end{cases}$$

where the quantity $l_j$ is defined recursively by

$$l_j = \begin{cases} 0, & j \leq 0, \\ \frac{1}{2^{f-1}}, & j = 1, \\ \frac{1}{2^{f-1}} \left[ \frac{1}{2} + l_{j-(f-1)} \right], & 2 \leq j \leq M - f + 2. \end{cases} \qquad (2.8)$$

## 2.17 Yamamoto Flag Code

Wang's code requires reversing the bits of the integer $n$ before it is encoded and reversing it again as the last step in its decoding. This code can be somewhat improved if we simply move the MSB of $n$ (which is always a 1) to its right end, to separate the codeword from the flag. The fact that the MSB of $n$ is included in the codeword of $n$ introduces a slight redundancy because this bit is known to be a 1. This bit is needed in Wang's code, because it helps to identify the flag at the end of a codeword. The flag code of this section, due to Hirosuke Yamamoto and Hiroshi Ochi [Yamamoto and Ochi 91], is more complex to understand and implement, but is faster to encode and decode and does not include the MSB of $n$. It is therefore slightly shorter on average. In addition, this code is a true flag code, because it is UD even though the bit pattern of the flag may appear inside a codeword. We first explain the encoding process in the simple case $f = 3$ (a 3-bit flag).

We start with a positive integer $n = b_0 b_1 b_2 \ldots b_M$ and the particular 3-bit flag $\mathbf{p} = p_1 p_2 p_3 = 100$. We ignore $b_0$ because it is a 1. Starting at $b_1$, we compare overlapping pairs of bits $b_i b_{i+1}$ to the fixed pair $p_1 p_2 = 10$. If $b_i b_{i+1} \neq 10$, we append $b_i$ to the codeword-so-far and continue with the pair $b_{i+1} b_{i+2}$. If, however, $b_i b_{i+1} = 10$, we append the triplet 101 (which is $b_i b_{i+1} \overline{p}_f$) to the codeword-so-far and continue with the pair $b_{i+2} b_{i+3}$. Notice that $b_i b_{i+1} \overline{p}_f = 101$ is different from the flag, a fact exploited by the decoder. At the end, when all the pairs $b_i b_{i+1}$ have been checked, the LSB $b_M$ of $n$ may be left over. This bit is appended to the codeword-so-far, and is followed by the flag.

We encode $n = 325 = 101000101_2$ as an example. Its nine bits are numbered $b_0 b_1 \ldots b_8$ with $b_0 = 1$. The codeword-so-far starts as an empty string. The first pair is $b_1 b_2 = 01 \neq 10$, so $b_1 = 0$ is appended to the codeword-so-far. The next pair is $b_2 b_3 = 10$, so the triplet 101 is appended to the codeword-so-far, which becomes 0|101. The next pair is $b_4 b_5 = 00 \neq 10$, so $b_4 = 0$ is appended to the codeword-so-far. The next pair is $b_5 b_6 = 01 \neq 10$, so $b_5 = 0$ is appended to the codeword-so-far, which becomes 0|101|0|0. The next pair is $b_6 b_7 = 10$, so the triplet 101 is appended to the codeword-so-far which becomes 0|101|0|0|101. Once $b_6$ and $b_7$ have been included in the codeword-so-far, only $b_8 = 1$ remains and it is simply appended, followed by the flag bits. The resulting codeword is 0|101|0|0|101|1|100 where the two bits in boldface are the complement of $p_3$. Notice that the flag pattern 100 also appears inside the codeword.

It is now easy to see how such a codeword can be decoded. The decoder initializes the decoded bitstring to 1 (the MSB of $n$). Given a string of bits $a_1 a_2 \ldots$ whose length is unknown, the decoder scans it, examining overlapping triplets of bits and looking for the pattern $a_i a_{i+1} \overline{p}_f = 101$. When such a triplet is found, its last bit is discarded and the first two bits 10 are appended to the decoded string. The process stops when the flag is located.

In our example, the decoder appends $a_1 = 0$ to the decoded string. It then locates $a_2 a_3 a_4 = 101$, discards $a_4$, and appends $a_3 a_4$ to the decoded string. Notice that the codeword contains the flag pattern 100 in bits $a_4 a_5 a_6$ but this pattern disappears once $a_4$ has been discarded. The only potential problem in decoding is that an $f$-bit pattern of the form $\ldots a_{j-1} a_j | p_1 p_2 \ldots$ (i.e., some bits from the end of the codeword, followed by some bits from the start of the flag) will equal the flag. This problem is solved by selecting the special flag 100. For the special case $f = 3$, it is easy to verify that bit patterns of the form $xx|1$ or $x|10$ cannot equal 100. In the general case, a flag of the form $10^{f-1}$ (a 1 followed by $f - 1$ zeros) is selected, and again it is easy to see that no $f$-bit string of the form $xx \ldots x|100 \ldots 0$ can equal the flag.

We are now ready to present the general case, where $f$ can have any value greater than 1. Given a positive integer $n = b_0 b_1 b_2 \ldots b_M$ and the $f$-bit flag $\mathbf{p} = p_1 p_2 \ldots p_f = 10^{f-1}$, the encoder initializes the codeword-so-far $C$ to the empty string and sets a counter $t$ to 1. It then performs the following loop:

1. If $t > M - f + 2$ then [if $t \leq M$, then $C \leftarrow C + b_t b_{t+1} \ldots b_M$ endif],
$$C \leftarrow C + p_1 p_2 \ldots p_f, \text{ Stop.}$$
   endif

2. If $b_t b_{t+1} \cdots b_{t+(f-2)} \neq p_1 p_2 \ldots p_{f-1}$
      then $[C \leftarrow C + b_t, \ t \leftarrow t + 1]$,
      else $[C \leftarrow C + b_t b_{t+1} \ldots b_{t+(f-2)} \overline{p_f}, \ t \leftarrow t + (f - 1)]$
   endif
   Go to step 1.

Step 1 should read: If no more tuples remain [if some bits remain, append them to $C$, endif], append the flag to $C$. Stop.

Step 2 should read: If a tuple (of $f - 1$ bits from $n$) does not equal the most-significant $f - 1$ bits of the flag, append the next bit $b_t$ to $C$ and increment $t$ by 1. Else, append the entire tuple to $C$, followed by the complement of $p_f$, and increment $t$ by $f - 1$. Endif. Go to step 1.

The developers of this code prove that the choice of the bit pattern $10^{f-1}$, or equivalently $01^{f-1}$, for the flag guarantees that no string of the form $xx \ldots x | p_1 p_2 \ldots p_j$

can equal the flag. They also propose flags of the form $1^2 0^{f-2}$ for cases where $f \geq 4$.

The decoder initializes the decoded string $D$ to 1 (the MSB of $n$) and a counter $s$ to 1. It then iterates the following step until it finds the flag and stops.

```
If  a_s a_{s+1} ... a_{s+(f-2)} ≠ p_1 p_2 ... p_{f-1}
  then  D ← D + a_s ,  s ← s + 1,
  else
   [if  a_{s+(f-1)} = p_f
      then Stop,
      else  D ← D + a_s a_{s+1} ... a_{s+(f-2)} ,  s ← s + f
   endif]
endif
```

The developers also show that this code is universal and is almost asymptotically optimal in the sense of Equation (2.7). The codeword length $L(n)$ increases with $n$, but not monotonically, and is bounded by

$$\lfloor \log_2 n \rfloor + f \leq L(n) \leq \lfloor \log_2 n \rfloor + \frac{\lfloor \log_2 n \rfloor}{f-1} + f \leq \frac{f}{f-1} \log_2 n + f.$$

In the special case where each bit $b_j$ of $n$ (except the MSB $b_0$, which is not used by this code) is selected independently with probability $P(b_j = 0) = P(b_j = 1) = 1/2$ for $j = 1, 2, \ldots, M$, the average codeword length $\overline{L(n)}$ depends on $M$ and $f$ in a simple way (compare with Equation (2.8))

$$\overline{L(n)} = \begin{cases} M + f & \text{if } M \leq f - 2, \\ M + \frac{M-f+2}{2^{f-1}} + f, & \text{if } M \geq f - 1. \end{cases} \tag{2.9}$$

Table 2.32 lists some of the Yamamoto codes for $f = 3$ and $f = 4$. These are compared with the similar $S(r+1, 01^r)$ codes of Capocelli (Section 2.20.1 and Table 2.37).

Recall that the encoder inserts the intercalary bit $\overline{p_f}$ whenever it finds the pattern $p_1 p_2 \ldots p_{f-1}$ in the integer $n$ that is being encoded. Thus, if $f$ is small (a short flag), large values of $n$ may be encoded into long codewords because of the many intercalary bits. On the other hand, large $f$ (a long flag) results in long codewords for small values of $n$ because the flag has to be appended to each codeword. This is why a scheme where the flag starts small and becomes longer with increasing $n$ seems ideal. Such a scheme, dubbed dynamically-variable-flag-length (DVFL), has been proposed by Yamamoto and Ochi as an extension of their original code.

The idea is to start with an initial flag length $f_0$ and increment it by 1 at certain points. A function $T(f)$ also has to be chosen that satisfies $T(f_0) \geq 1$ and $T(f + 1) - T(f) \geq f - 1$ for $f \geq f_0$. Given a large integer $n = b_0 b_1 \ldots b_M$, the encoder (and also the decoder, working in lockstep) will increment $f$ when it reaches bits whose indexes equal $T(f_0)$, $T(f_0 + 1)$, $T(f_0 + 2)$, and so on. Thus, bits $b_1 b_2 \ldots b_{T(f_0)}$ of $n$ will be encoded with a flag length $f_0$, bits $b_{T(f_0)+1} b_{T(f_0)+2} \ldots b_{T(f_0+1)}$ will be encoded with a flag length $f_0 + 1$, and so on. In the original version, the encoder maintains a counter $t$ and examines the $f - 1$ bits starting at bit $b_t$. These are compared to the first $f - 1$ bits $10^{f-2}$ of the flag. In the extended version, $f$ is determined by the counter $t$ and

| $n$ | $S(3,011)$ | $Y_3(n)$ | $S(4,0111)$ | $Y_4(n)$ |
|---|---|---|---|---|
| 1 | 011 | 011 | 0111 | 0111 |
| 2 | 0 011 | 0 011 | 0 0111 | 0 0111 |
| 3 | 1 011 | 1 011 | 1 0111 | 1 0111 |
| 4 | 00 011 | 00 011 | 00 0111 | 00 0111 |
| 5 | 01 011 | 010 011 | 01 0111 | 01 0111 |
| 6 | 10 011 | 10 011 | 10 0111 | 10 0111 |
| 7 | 11 011 | 11 011 | 11 0111 | 11 0111 |
| 8 | 000 011 | 000 011 | 000 0111 | 000 0111 |
| 9 | 001 011 | 0010 011 | 001 0111 | 001 0111 |
| 10 | 010 011 | 0100 011 | 010 0111 | 010 0111 |
| 11 | 100 011 | 0101 011 | 011 0111 | 0110 0111 |
| 12 | 101 011 | 100 011 | 100 0111 | 100 0111 |
| 13 | 110 011 | 1010 011 | 101 0111 | 101 0111 |
| 14 | 111 011 | 110 011 | 110 0111 | 110 0111 |
| 15 | 0000 011 | 111 011 | 111 0111 | 111 0111 |
| 16 | 0001 011 | 0000 011 | 0000 0111 | 0000 0111 |
| 17 | 0010 011 | 00010 011 | 0001 0111 | 0001 0111 |
| 18 | 0100 011 | 00100 011 | 0010 0111 | 0010 0111 |
| 19 | 0101 011 | 00101 011 | 0011 0111 | 00110 0111 |
| 20 | 1000 011 | 01000 011 | 0100 0111 | 0100 0111 |
| 21 | 1001 011 | 010010 011 | 0101 0111 | 0101 0111 |
| 22 | 1010 011 | 01010 011 | 0110 0111 | 01100 0111 |
| 23 | 1100 011 | 01011 011 | 1000 0111 | 01101 0111 |
| 24 | 1101 011 | 1000 011 | 1001 0111 | 1000 0111 |
| 25 | 1110 011 | 10010 011 | 1010 0111 | 1001 0111 |
| 26 | 1111 011 | 10100 011 | 1011 0111 | 1010 0111 |
| 27 | 00000 011 | 10101 011 | 1100 0111 | 10110 0111 |
| 28 | 00001 011 | 1100 011 | 1101 0111 | 1100 0111 |
| 29 | 00010 011 | 11010 011 | 1110 0111 | 1101 0111 |
| 30 | 00100 011 | 1110 011 | 1111 0111 | 1110 0111 |
| 31 | 00101 011 | 1111 011 | 00000 0111 | 1111 0111 |
| 32 | 01000 011 | 00000 011 | 00001 0111 | 00000 0111 |

Table 2.32: Some Capocelli and Yamamoto Codes for $f = 3$ and $f = 4$.

the function $T$ by solving the inequality $T(f-1) < t \le T(f)$. In the last step, the flag $10^{f_M - 1}$ is appended to the codeword, where $f_M$ is determined by the inequality $T(f_M - 1) < M + 1 \le T(f_M)$. The encoding steps are as follows:

1. Initialize $f \leftarrow f_0$, $t \leftarrow 1$, and the codeword $C$ to the empty string.

2. If $t > M - f + 2$
   then [if $t \le M$ then $C \leftarrow C + b_t b_{t+1} \ldots b_M$ endif],
   [if $M + 1 > T(f)$ then $f \leftarrow f + 1$ endif],
   $C \leftarrow C + 10^{f-1}$, Stop.
endif.

3. If $b_t b_{t+1} \ldots b_{t+(f-2)} \neq 10^{f-2}$
     then $C \leftarrow C + b_t,\ t \leftarrow t + 1,$
          [if $t > T(f)$ then $f \leftarrow f + 1$ endif]
     else $C \leftarrow C + b_t b_{t+1} \ldots b_{t+(f-2)} 1,$
          $t \leftarrow t + (f - 1),$
          [if $t > T(f)$ then $f \leftarrow f + 1$ endif]
   endif, Go to step 2.

The decoding steps are the following:

1. Initialize $f \leftarrow f_0,\ t \leftarrow 1,\ s \leftarrow 1,\ D \leftarrow 1.$

2. If $a_s a_{s+1} \ldots a_{s+(f-2)} \neq 10^{f-2}$
     then $D \leftarrow D + a_s,\ s \leftarrow s + 1,\ t \leftarrow t + 1,$
          [if $t > T(f)$ then $f \leftarrow f + 1$ endif],
     else [if $a_{s+(f-1)} = 0$ then Stop
                              else
   $D \leftarrow D + a_s a_{s+1} \ldots a_{s+(f-2)},\ s \leftarrow s + f,$
   $t \leftarrow t + (f - 1),$ [if $t > T(f)$ then $f \leftarrow f + 1$ endif]
              endif]
     endif, Go to step 2.

The choice of function $T(f)$ is the next issue. The authors show that a function of the form $T(f) = Kf(f - 1)$, where $K$ is a nonnegative constant, results in an asymptotically optimal DVFL. There are other functions that guarantee the same property of DVFL, but the point is that varying $f$, while generating short codewords, also eliminates one of the chief advantages of any flag code, namely its robustness. If the length $f$ of the flag is increased during the construction of a codeword, then any future error may cause the decoder to lose synchronization and may propagate indefinitely through the string of codewords. It seems that, in practice, the increased reliability achieved by synchronization with a fixed $f$ overshadows the savings produced by a dynamic encoding scheme that varies $f$.

## 2.18 Number Bases

This short section is a digression to prepare the reader for the Fibonacci and Goldbach codes that follow. Decimal numbers use base 10. The number $2037_{10}$, for example, has a value of $2 \times 10^3 + 0 \times 10^2 + 3 \times 10^1 + 7 \times 10^0$. We can say that 2037 is the sum of the digits 2, 0, 3, and 7, each weighted by a power of 10. Fractions are represented in the same way, using negative powers of 10. Thus, $0.82 = 8 \times 10^{-1} + 2 \times 10^{-2}$ and $300.7 = 3 \times 10^2 + 7 \times 10^{-1}$.

Binary numbers use base 2. Such a number is represented as a sum of its digits, each weighted by a power of 2. Thus, $101.11_2 = 1 \times 2^2 + 0 \times 2^1 + 1 \times 2^0 + 1 \times 2^{-1} + 1 \times 2^{-2}$.

Since there is nothing special about 10 or 2 (actually there is, because 2 is the smallest integer that can be a base for a number system and 10 is the number of our fingers), it should be easy to convince ourselves that any positive integer $n > 1$ can serve as the basis for representing numbers. Such a representation requires $n$ "digits" (if $n \geq 10$, we use the ten digits and the letters A, B, C, ...) and represents the number $d_3 d_2 d_1 d_0 . d_{-1}$ as the sum of the digits $d_i$, each multiplied by a power of $n$, thus $d_3 n^3 + d_2 n^2 + d_1 n^1 + d_0 n^0 + d_{-1} n^{-1}$. The base of a number system does not have to consist of powers of an integer but can be any *superadditive* sequence that starts with 1.

Definition: A superadditive sequence $a_0, a_1, a_2, \ldots$ is one where any element $a_i$ is greater than the sum of all its predecessors. An example is $1, 2, 4, 8, 16, 32, 64, \ldots$ where each element equals 1 plus the sum of all its predecessors. This sequence consists of the familiar powers of 2, so we know that any integer can be expressed by it using just the digits 0 and 1 (the two bits). Another example is $1, 3, 6, 12, 24, 50, \ldots$, where each element equals 2 plus the sum of all its predecessors. It is easy to see that any integer can be expressed by it using just the digits 0, 1, and 2 (the three trits).

Given a positive integer $k$, the sequence $1, 1 + k, 2 + 2k, 4 + 4k, \ldots, 2^i(1 + k)$ is superadditive, because each element equals the sum of all its predecessors plus $k$. Any nonnegative integer can be represented uniquely in such a system as a number $x \ldots xxy$, where $x$ are bits and $y$ is a single digit in the range $[0, k]$.

In contrast, a general superadditive sequence, such as 1, 8, 50, 3102 can be used to represent integers, but not uniquely. The number 50, e.g., equals $8 \times 6 + 1 + 1$, so it can be represented as $0062 = 0 \times 3102 + 0 \times 50 + 6 \times 8 + 2 \times 1$, but also as $0100 = 0 \times 3102 + 1 \times 50 + 0 \times 8 + 0 \times 1$.

It can be shown that $1 + r + r^2 + \cdots + r^k$ is less than $r^{k+1}$ for any real number $r > 1$. This implies that the powers of any real number $r > 1$ can serve as the base of a number system using the digits $0, 1, 2, \ldots, d$ for some $d < r$.

The number $\phi = \frac{1}{2}(1 + \sqrt{5}) \approx 1.618$ is the well-known golden ratio. It can serve as the base of a number system, with 0 and 1 as the digits. Thus, for example, $100.1_\phi = \phi^2 + \phi^{-1} \approx 3.23_{10}$.

Some real bases have special properties. For example, any positive integer $R$ can be expressed as $R = b_1 F_1 + b_2 F_2 + b_3 F_3 + b_4 F_5 + b_5 F_8 + b_6 F_{13} \cdots$, where $b_i$ are either 0 or 1, and $F_i$ are the Fibonacci numbers $1, 2, 3, 5, 8, 13, \ldots$. This representation has the interesting property, known as Zeckendorf's theorem [Zeckendorf 72], that the string $b_1 b_2 \ldots$ does not contain any adjacent 1's. This useful property, which is the basis for the Goldbach codes (Section 2.21) is easy to prove. If an integer $I$ in this representation has the form $\ldots 01100 \ldots$, then because of the definition of the Fibonacci numbers, $I$ can be written $\ldots 00010 \ldots$.

Examples are the integer 5, whose Fibonacci representation is 0001 and $33 = 1 + 3 + 8 + 21$, which is expressed in the Fibonacci base as the 7-bit number 1010101. Section 2.15 discusses the $n$-step Fibonacci numbers, defined by Equation (2.6).

The Australian Aboriginals use a number of languages, some of which employ binary or binary-like counting systems. For example, in the Kala Lagaw Ya language, the numbers 1 through 6 are urapon, ukasar, ukasar-urapon, ukasar-ukasar, ukasar-ukasar-urapon, and ukasar-ukasar-ukasar.

The familiar terms "dozen" (12) and "gross" (twelve dozen) originated in old duodecimal (base 12) systems of measurement.

Computers use binary numbers mostly because it is easy to design electronic circuits with just two states, as opposed to ten states.

### Leonardo Pisano Fibonacci (1170–1250)

Leonard of Pisa (or Fibonacci), was an Italian mathematician, often considered the greatest mathematician of the Middle Ages. He played an important role in reviving ancient mathematics and also made significant original contributions. His book, *Liber Abaci*, introduced the modern decimal system and the use of Arabic numerals and the zero into Europe.

Leonardo was born in Pisa around 1170 or 1180 (he died in 1250). His father Guglielmo was nicknamed Bonaccio (the good natured or simple), which is why Leonardo is known today by the nickname Fibonacci (derived from filius Bonacci, son of Bonaccio). He is also known as Leonardo Pisano, Leonardo Bigollo, Leonardi Bigolli Pisani, Leonardo Bonacci, and Leonardo Fibonacci.

The father was a merchant (and perhaps also the consul for Pisa) in Bugia, a port east of Algiers (now Bejaïa, Algeria), and Leonardo visited him there while still a boy. It seems that this was where he learned about the Arabic numeral system.

Realizing that representing numbers with Arabic numerals rather than with Roman numerals is easier and greatly simplifies the arithmetic operations, Fibonacci traveled throughout the Mediterranean region to study under the leading Arab mathematicians of the time. Around 1200 he returned to Pisa, where in 1202, at age 32, he published what he had learned in *Liber Abaci*, or Book of Calculation.

Leonardo became a guest of the Emperor Frederick II, who enjoyed mathematics and science. In 1240, the Republic of Pisa honored Leonardo, under his alternative name of Leonardo Bigollo, by granting him a salary.

Today, the Fibonacci sequence is one of the best known mathematical objects. This sequence and its connection to the golden ratio $\phi = \frac{1}{2}(1 + \sqrt{5}) \approx 1.618$ have been studied extensively and the mathematics journal *Fibonacci Quarterly* is dedicated to the Fibonacci and similar sequences. The publication [Grimm 73] is a short biography of Fibonacci.

## 2.19 Fibonacci Code

The Fibonacci code, as its name suggests, is closely related to the Fibonacci representation of the integers. The Fibonacci code of the positive integer $n$ is the Fibonacci representation of $n$ with an additional 1 appended to the right end. Thus, the Fibonacci code of 5 is 0001|1 and that of 33 is 1010101|1. It is obvious that such a code ends with a pair 11, and that this is the only such pair in the codeword (because the Fibonacci representation does not have adjacent 1's). This property makes it possible to decode such a code unambiguously, but also causes these codes to be long, because not having adjacent 1's restricts the number of possible binary patterns.

| 1 | 11 | 7 | 01011 |
|---|---|---|---|
| 2 | 011 | 8 | 000011 |
| 3 | 0011 | 9 | 100011 |
| 4 | 1011 | 10 | 010011 |
| 5 | 00011 | 11 | 001011 |
| 6 | 10011 | 12 | 101011 |

Table 2.33: Twelve Fibonacci Codes.

Table 2.33 lists the first 12 Fibonacci codes.

**Decoding.** Skip bits of the code until a pair 11 is reached. Replace this 11 by 1. Multiply the skipped bits by the values ..., 13, 8, 5, 3, 2, 1 (the Fibonacci numbers), and add the products. Obviously, it is not necessary to do any multiplication. Simply use the 1 bits to select the proper Fibonacci numbers and add.

The Fibonacci codes are long, but have the advantage of being more robust than most other variable-length codes. A corrupt bit in a Fibonacci code may change a pair of consecutive bits from 01 or 10 to 11 or from 11 to 10 or 01. In the former case, a code may be read as two codes, while in the latter case two codes will be decoded as a single code. In either case, the slippage will be confined to just one or two codewords and will not propagate further.

The length of the Fibonacci code for $n$ is less than or equal to $1 + \lfloor \log_\phi \sqrt{5}n \rfloor$ where $\phi$ is the golden ratio (compare with Equation (1.1)).

Figure 2.34 represents the lengths of two Elias codes and the Fibonacci code graphically and compares them to the length of the standard binary (beta) code.

In [Fraenkel and Klein 96] (and also [Fraenkel and Klein 85]), the authors denote this code by $C^1$ and show that it is universal, with $c1 = 2$ and $c2 = 3$, but is not asymptotically optimal because $c1 > 1$. They also prove that for any length $r \geq 1$, there are $F_r$ codewords of length $r + 1$ in $C^1$. As a result, the total number of codewords of length up to and including $r$ is $\sum_{i=1}^{r-1} F_i = F_{r+1} - 1$. (See also Figure 2.41.)

A P-code is a set of codewords that end with the same pattern $P$ (the pattern is the suffix of the codewords) and where no codeword includes the pattern anywhere else. Given a $k$-bit binary pattern $P$, the set of all binary strings of length $\geq k$ in which $P$ occurs only as a suffix is called the set generated by $P$ and is denoted by $\mathcal{L}(P)$. In [Berstel and Perrin 85] this set is called a semaphore code. All codewords in the $C^1$ code end with the pattern $P = 11$, so this code is $\mathcal{L}(11)$.

The next Fibonacci code proposed by Fraenkel and Klein is denoted by $C^2$ and is constructed from $C^1$ as follows:

1. Each codeword in $C^1$ ends with two consecutive 1's; delete one of them.
2. Delete all the codewords that start with 0.

Thus, the first few $C^2$ codewords, constructed with the help of Table 2.33, are 1, 101, 1001, 10001, 10101, 100001, and 101001. An equivalent procedure to construct this code is the following:

1. Delete the rightmost 1 of every codeword in $C^1$.
2. Prepend 10 to every codeword.
3. Include 1 as the first codeword.

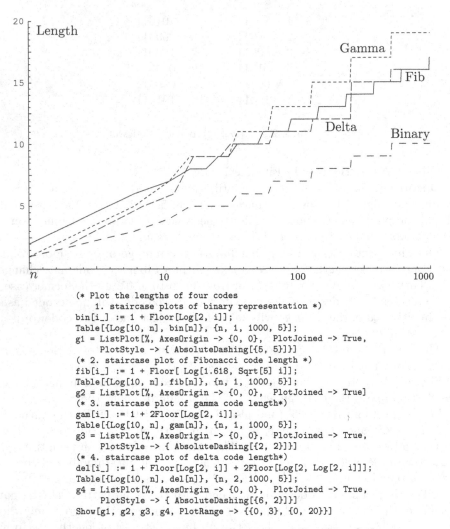

```
(* Plot the lengths of four codes
   1. staircase plots of binary representation *)
bin[i_] := 1 + Floor[Log[2, i]];
Table[{Log[10, n], bin[n]}, {n, 1, 1000, 5}];
g1 = ListPlot[%, AxesOrigin -> {0, 0}, PlotJoined -> True,
     PlotStyle -> { AbsoluteDashing[{5, 5}]}]
(* 2. staircase plot of Fibonacci code length *)
fib[i_] := 1 + Floor[ Log[1.618, Sqrt[5] i]];
Table[{Log[10, n], fib[n]}, {n, 1, 1000, 5}];
g2 = ListPlot[%, AxesOrigin -> {0, 0}, PlotJoined -> True]
(* 3. staircase plot of gamma code length*)
gam[i_] := 1 + 2Floor[Log[2, i]];
Table[{Log[10, n], gam[n]}, {n, 1, 1000, 5}];
g3 = ListPlot[%, AxesOrigin -> {0, 0}, PlotJoined -> True,
     PlotStyle -> { AbsoluteDashing[{2, 2}]}]
(* 4. staircase plot of delta code length*)
del[i_] := 1 + Floor[Log[2, i]] + 2Floor[Log[2, Log[2, i]]];
Table[{Log[10, n], del[n]}, {n, 2, 1000, 5}];
g4 = ListPlot[%, AxesOrigin -> {0, 0}, PlotJoined -> True,
     PlotStyle -> { AbsoluteDashing[{6, 2}]}]
Show[g1, g2, g3, g4, PlotRange -> {{0, 3}, {0, 20}}]
```

Figure 2.34: Lengths of Binary, Fibonacci and Two Elias Codes.

A simple check verifies the equivalence of the two constructions. Code $C^2$ is not a straightforward Fibonacci code as is $C^1$, but it can be termed a Fibonacci code, because the interior bits of each codeword correspond to Fibonacci numbers. The code consists of one codeword of length 1, no codewords of length 2, and $F_{r-2}$ codewords of length $r$ for any $r \geq 3$.

The $C^2$ code is not a prefix code, but is UD. Individual codewords are identified by the decoder because each starts and ends with a 1. Thus, two consecutive 1's indicate the boundary between codewords. The first codeword introduces a slight complication, but can be handled by the decoder. A string of the form $\dots 011110 \dots$ is interpreted by the decoder as $\dots 01|1|1|10 \dots$, i.e., two consecutive occurrences of the codeword 1.

The $C^2$ code is also more robust than $C^1$. A single error cannot propagate far because the decoder is looking for the pattern 11. The worst case is a string of the form $\ldots xyz \ldots = \ldots 1|10\ldots01|1|10\ldots01|1\ldots$ where the middle 1 gets corrupted to a 0. This results in $\ldots 1|10\ldots01010\ldots01|1\ldots$ which is interpreted by the decoder as one long codeword. The three original codewords $xyz$ are lost, but the error does not propagate any further. Other single errors (continued or lost first result in the loss of only two codewords.

The third Fibonacci code described in [Fraenkel and Klein 96] is denoted by $C^3$ and is constructed from $C^1$ as follows:

1. Delete the rightmost 1 of every codeword of $C^1$.
2. For every $r \geq 1$, select the set of $C^1$ codewords of length $r$, duplicate the set, and distinguish between the two copies by prepending a 10 to all the codewords in one copy and a 11 to all the codewords in the other copy.

This results in the codewords 101, 111, 1001, 1101, 10001, 10101, 11001, 11101, 100001, 101001, 100101, 110001, .... It is easy to see that every codeword of $C^3$ starts with a 1, has at most three consecutive 1's (and then they appear in the prefix), and every codeword except the second ends with 01. The authors show that for any $r \geq 3$ there are $2F_{r-2}$ codewords. It is easy to see that $C^3$ is not a prefix code because, for example, codeword 111 is also the prefix of 11101. However, the code is UD. The decoder first checks for the pattern 111 and interprets it depending on the bit that follows. If that bit is 0, then this is a codeword that starts with 111; otherwise, this is the codeword 111 followed by another codeword. If the current pattern is not 111, the decoder checks for 011. Every codeword except 111 ends with 01, so the pattern 01|1 indicates the end of a codeword and the start of the next one. This pattern does not appear inside any codeword.

Given an $r$-bit codeword $y_1 y_2 \ldots y_r$ (where $y_r = 1$), the developers show that its index (i.e., the integer whose codeword it is) is given by

$$2F_{r-1} - 2 + y_2 F_{r-2} + \sum_{i=3}^{r} y_i F_{i-1} - F_{r-1} + 1$$

$$= \sum_{i=3}^{r+2} y_i F_{i-1} + (y_2 - 1)F_{r-2} - 1,$$

where $y_{r+1} = 1$ is the leftmost bit of the next codeword.

The developers compare the three Fibonacci codes with the Huffman code using English text of 100,000 words collected from many different sources. Letter distributions were computed and were used to assign Huffman, $C^1$, $C^2$, and $C^3$ codes to the 26 letters. The average lengths of the codes were 4.185, 4.895, 5.298, and 4.891 bits/character, respectively. The Huffman code has the shortest length, but is vulnerable to storage and transmission errors. The conclusion is that the three Fibonacci codes are good candidates for compressing data in applications where robustness is more important than optimal compression.

It is interesting to note that the Fibonacci sequence can be generalized by adding a parameter. An $m$-step Fibonacci number equals the sum of its $m$ immediate predecessors (Equation (2.6)). Thus, the common Fibonacci numbers are 2-step. The $m$-step Fibonacci numbers can be used to construct order-$m$ Fibonacci codes (Section 2.20), but none of these codes is asymptotically optimal.

The anti-Fibonacci numbers are defined recursively as $f(1) = 1$, $f(2) = 0$, and $f(k + 2) = f(k) - f(k + 1)$. The sequence starts with 1, 0, 1, $-1$, 2, $-3$, 5, $-8$, .... This sequence is also obtained when the Fibonacci sequence is extended backward from 0. Thus,

$$\ldots - 8, 5, -3, 2, -1, 1, 0, 1, 1, 2, 3, 5, 8, \ldots.$$

In much the same way that the ratios of successive Fibonacci numbers converge to $\phi$, the ratios of successive anti-Fibonacci numbers converge to $1/\phi$.

## 2.20 Generalized Fibonacci Codes

The Fibonacci code of Section 2.19 is elegant and robust. The generalized Fibonacci codes presented here, due to Alberto Apostolico and Aviezri Fraenkel [Apostolico and Fraenkel 87], are also elegant, robust, and UD. They are based on $m$-step Fibonacci numbers (sometimes also called generalized Fibonacci numbers). The authors show that these codes are easy to encode and decode and can be employed to code integers as well as arbitrary, unbound bit strings. (The MSB of an integer is 1, but the MSB of an arbitrary string can be any bit.)

The sequence $F^{(m)}$ of $m$-step Fibonacci numbers $F_n^{(m)}$ is defined as follows:

$$F_n^{(m)} = F_{n-1}^{(m)} + F_{n-2}^{(m)} + \cdots + F_{n-m}^{(m)}, \text{ for } n \geq 1,$$

and

$$F_{-m+1}^{(m)} = F_{-m+2}^{(m)} = \cdots = F_{-2}^{(m)} = 0, \quad F_{-1}^{(m)} = F_0^{(m)} = 1.$$

Thus, for example, $F_1^{(m)} = 2$ for any $m$, while for $m = 3$ we have $F_{-2}^{(3)} = 0$, $F_{-1}^{(3)} = F_0^{(3)} = 1$, which implies $F_1^{(3)} = F_0^{(3)} + F_{-1}^{(3)} + F_{-2}^{(3)} = 2$, $F_2^{(3)} = F_1^{(3)} + F_0^{(3)} + F_{-1}^{(3)} = 4$, and $F_3^{(3)} = F_2^{(3)} + F_1^{(3)} + F_0^{(3)} = 7$.

The generalized Fibonacci numbers can serve as the basis of a numbering system. Any positive integer $N$ can be represented as the sum of several distinct $m$-step Fibonacci numbers and this sum does not contain $m$ consecutive 1's. Thus, if we represent $N$ in this number basis, the representation will not have a run of $m$ consecutive 1's. An obvious conclusion is that such a run can serve as a comma, to separate consecutive codewords.

The two generalized Fibonacci codes proposed by Apostolico and Fraenkel are pattern codes (P-codes) and are also universal and complete (a UD code is complete if the addition of any codeword turns it into a non-UD code). A P-code is a set of codewords that end with the same pattern $P$ (the pattern is the suffix of the codewords) and where no codeword includes the pattern anywhere else. For example, if $P = 111$, then a P-code is a set of codewords that all end with 111 and none has 111 anywhere else. A P-code is a prefix code because once a codeword $c = x \ldots x111$ has been selected, no other codeword will start with $c$ because no other codeword has 111 other than as a suffix.

Given a P-code with the general pattern $P = a_1 a_2 \ldots a_p$ and an arbitrary codeword $x_1 x_2 \ldots x_n a_1 a_2 \ldots a_p$, we consider the string $a_2 a_3 \ldots a_p | x_1 x_2 \ldots x_n a_1 a_2 \ldots a_{p-1}$ (the end of $P$ followed by the start of a codeword). If $P$ happens to appear anywhere in this string, then the code is not synchronous and a transmission error may propagate over many codewords. On the other hand, if $P$ does not appear anywhere inside such a string, the code is synchronous and is referred to as an SP-code (It is also sometimes robust). Such codes are useful in applications where data integrity is important. When an error occurs during storage or transmission of an encoded message, the decoder loses synchronization, but regains it when it sees the next occurrence of $P$ (and it sees it when it reads the next codeword). As usual, there is a price to pay for this useful property. In general, SP-codes are not complete.

The authors show that the Elias delta code is asymptotically longer than the generalized Fibonacci codes (of any order) for small integers, but becomes shorter at a certain point that depends on the order $m$ of the generalized code. For $m = 2$, this transition point is at $F_{27}^{(2)} - 1 = 514, 228$. For $m = 3$ it is at $(F_{63}^{(3)} + F_{631}^{(3)} - 1)/2 = 34, 696, 689, 675, 849, 696 \approx 3.47 \times 10^{16}$, and for $m = 4$ it is at $(F_{231}^{(4)} + F_{229}^{(4)} + F_{228}^{(4)} - 1)/3 \approx 4.194 \times 10^{65}$.

The first generalized Fibonacci code, $C_1^{(m)}$, employs the pattern $1^m$. The code is constructed as follows:

1. The $C_1^{(m)}$ code of $N = 1$ is $1^m$ and the $C_1^{(m)}$ code of $N = 2$ is $01^m$.

2. All other $C_1^{(m)}$ codes have the suffix $01^m$ and a prefix of $n-1$ bits, where $n$ starts at 2. For each $n$, there are $F_n^{(m)}$ prefixes which are the $(n-1)$-bit $F^{(m)}$ representations of the integers 0, 1, 2, ....

Table 2.35 lists examples of $C_1^{(3)}$. Once the codes of $N = 1$ and $N = 2$ are in place, we set $n = 2$ and construct the 1-bit $F^{(3)}$ representations of the integers 0, 1, 2, .... There are only two such representations, 0 and 1, and they become the prefixes of the next two codewords (for $N = 3$ and $N = 4$). We increment $n$ to 3, and construct the 2-bit $F^{(3)}$ representations of the integers 0, 1, 2, .... There are four of them, namely 0, 1, 10, and 11. Each is extended to two bits to form 00, 01, 10, and 11, and they become the prefixes of $N = 5$ through $N = 8$.

Notice that $N_1 < N_2$ implies that the $C_1$ code of $N_1$ is lexicographically smaller than the $C_1$ code of $N_2$ (where a blank space is considered lexicographically smaller than 0). The $C_1$ code is a P-code (with $1^m$ as the pattern) and is therefore a prefix code. The code is partitioned into groups of $m + n$ codewords each. In each group, the prefix of the codewords is $n - 1$ bits long and the suffix is $01^m$. Each group consists of $F_{n-1}^{(m)}$ codewords of which $F_{n-2}^{(m)}$ codewords start with 0, $F_{n-3}^{(m)}$ start with 10, $F_{n-4}^{(m)}$ start with 110, and so on, up to $F_{n-m-1}^{(m)}$ codewords that start with $1^{m-1}0$.

A useful, interesting property of this code is that it can be used for arbitrary bitstrings, not just for integers. The difference is that an integer has a MSB of 1, whereas a binary string can start with any bit. The idea is to divide the set of bitstrings into those strings that start with 1 (integers) and those that start with 0. The former strings are assigned $C_1$ codes that start with 1 (in Table 2.35, those are the codes of 4, 7, 8, 13, 14, 15, ...). The latter are assigned $C_1$ codes that start with 0 in the following way. Given a bitstring of the form $0^i N$ (where $N$ is a nonnegative integer), assign to it the code $0^i N 01^m$. Thus, the string 001 is assigned the code 00|1|0111 (which happens to be the code of 10 in the table).

| String $M$ | $C_1^{(3)}$ 7421\|0111 | $F^{(3)}$ $\frac{1}{3}$7421 | $N$ |
|---|---|---|---|
|  |  | 0 | 0 |
| 0 | 111 | 1 | 1 |
| 00 | 0111 | 10 | 2 |
| 000 | 0\|0111 | 11 | 3 |
| 1 | 1\|0111 | 100 | 4 |
| 0000 | 00\|0111 | 101 | 5 |
| 01 | 01\|0111 | 110 | 6 |
| 2 | 10\|0111 | 1000 | 7 |
| 3 | 11\|0111 | 1001 | 8 |
| 00000 | 000\|0111 | 1010 | 9 |
| 001 | 001\|0111 | 1011 | 10 |
| 02 | 010\|0111 | 1100 | 11 |
| 03 | 011\|0111 | 1101 | 12 |
| 4 | 100\|0111 | 10000 | 13 |
| 5 | 101\|0111 | 10001 | 14 |
| 6 | 110\|0111 | 10010 | 15 |
| 000000 | 0000\|0111 | 10011 | 16 |

| $C_2^{(3)}$ $\frac{1}{3}$7421\|011 | $C_2^{(2)}$ $\frac{1}{3}$85321\|01 | $N$ |
|---|---|---|
| 11 | 1 | 1 |
| 1\|011 | 1\|01 | 2 |
| 10\|011 | 10\|01 | 3 |
| 11\|011 | 100\|01 | 4 |
| 100\|011 | 101\|01 | 5 |
| 101\|011 | 1000\|01 | 6 |
| 110\|011 | 1001\|01 | 7 |
| 1000\|011 | 1010\|01 | 8 |
| 1001\|011 | 10000\|01 | 9 |
| 1010\|011 | 10001\|01 | 10 |
| 1011\|011 | 10010\|01 | 11 |
| 1100\|011 | 10100\|01 | 12 |
| 1101\|011 | 10101\|01 | 13 |
| 10000\|011 | 100000\|01 | 14 |
| 10001\|011 | 100001\|01 | 15 |
| 10010\|011 | 100010\|01 | 16 |

Table 2.35: The Generalized Fibonacci Code $C_1^{(3)}$.    Table 2.36: Generalized Codes $C_2^{(3)}$ $C_2^{(2)}$.

The second generalized Fibonacci code, $C_2^{(m)}$, employs the pattern $1^{m-1}$. The code is constructed as follows:

1. The $C_2^{(m)}$ code of $N = 1$ is $1^{m-1}$.

2. All other $C_2^{(m)}$ codes have the suffix $01^{m-1}$ and prefixes that are the $F^{(m)}$ representations of the positive integers, in increasing order.

Table 2.36 shows examples of $C_2^{(3)}$ and $C_2^{(2)}$. Once the code of $N = 1$ has been constructed, we prepare the $F^{(m)}$ representation of $n = 1$ and prepend it to $01^{m-1}$ to obtain the $C_2$ code of $N = 2$. We then increment $n$ by 1, and construct the other codes in the same way. Notice that $C_2$ is not a prefix code because, for example, $C_2^{(3)}(2) = 1011$ is a prefix of $C_2^{(3)}(11) = 1011011$. However, $C_2$ is a UD code because the string $01^{m-1}|1$ separates any two codewords (each ends with $01^{m-1}$ and starts with 1). The $C_2$ codes become longer with $N$ and are organized in length groups for $n = 0, 1, 2, \ldots$. Each group has $F_{n-1}^{(m)} - F_{n-2}^{(m)}$ codewords of length $m + n - 1$ that can be partitioned as follows: $F_{n-3}^{(m)}$ codewords that start with 10, $F_{n-4}^{(m)}$ codewords that start with 110, and so on, up to $F_{n-m-1}^{(m)}$ codewords that start with $1^{m-1}0$.

The authors provide simple procedures for encoding and decoding the two codes. Once a value for $m$ has been selected, the procedures require tables of many $m$-step Fibonacci numbers (if even larger numbers are needed, they have to be computed on the fly).

A note on robustness. It seems that a P-code is robust. Following an error, the decoder will locate the pattern $P$ very quickly and will resynchronize itself. However, the term "robust" is imprecise and at least code $C_1^{(3)}$ has a weak point, namely the codeword 111. In his short communication [Capocelli 89], Renato Capocelli points out a case where the decoder of this code can be thrown completely off track because of this codeword. The example is the message $41^n3$, which is encoded in $C_1^{(3)}$ as $10111|(111)^n|00111$. If the second MSB becomes 1 because of an error, the decoder will not sense any error and will decode this string as $(111)^{n+1}|1100111$, which is the message $1^{n+1}(15)$.

## 2.20.1 A Related Code

The simple code of this section, proposed by Renato Capocelli [Capocelli 89], is prefix, complete, universal, and also synchronizable (see also Section 3.3 for more synchronous codes). It is not a generalized Fibonacci code, but it is related to the $C_1$ and $C_2$ codes of Apostolico and Fraenkel. The code depends on a parameter $r$ and is denoted by $S(r + 1, 01^r)$. Once $r$ has been selected, two-part codewords are constructed with a suffix $01^r$ and a prefix that is a binary string that does not contain the suffix. Thus, for example, if $r = 2$, the suffix is 011 and the prefixes are all the binary strings, starting with the empty string, that do not contain 011. Table 2.37 lists a few examples of $S(3, 011)$ (see also Table 2.32) and it is easy to see how the prefixes are the empty string, 0, 1, 00, 01, and so on, but they include no strings with 011. The codeword of $N = 9$ is $010|011$, but the codeword of $N = 10$ has the prefix 100 and not 011, so it is $100|011$. In general a codeword $x\ldots x0101y\ldots y|011$ will be followed by $x\ldots x1000y\ldots y|011$ instead of by $x\ldots x0110y\ldots y|011$. Such codewords have either the form $0\beta|011$ (where $\beta$ does not contain two consecutive 1's) or the form $1\gamma|011$ (where $\gamma$ does not contain 011). For example, only 12 of the 16 4-bit prefixes can be used by this code, because the four prefixes 0011, 0110, 0111, and 1011 contain the pattern 011. In general, the number of codewords of length $N + 3$ in $S(3, 011)$ is $F_{N+3} - 1$. For $N = 4$ (codewords of a 4-bit prefix and a 3-bit suffix), the number of codewords is $F_{4+3} - 1 = F_7 - 1 = 12$.

| $N$ | $S(3, 011)$ | BS |
|---|---|---|
| 0 | 011 | 0 |
| 1 | 0011 | 1 |
| 2 | 1011 | 00 |
| 3 | 00011 | 01 |
| 4 | 01011 | 10 |
| 5 | 10011 | 11 |
| 6 | 11011 | 000 |
| 7 | 000011 | 001 |
| 8 | 001011 | 010 |
| 9 | 010011 | 011 |
| 10 | 100011 | 100 |
| 11 | 101011 | 101 |
| 12 | 110011 | 110 |
| 13 | 111011 | 111 |

Table 2.37: Code $S(3, 011)$ for Integers $N$ and Strings BS.

The table also illustrates how $S(3,011)$ can be employed to encode arbitrary bit-strings, not just integers. Simply write all the binary strings in lexicographic order and assign them the $S(3,011)$ codewords in increasing order.

The special suffix $01^r$ results in synchronized codewords. If the decoder loses synchronization, it regains it as soon as it recognizes the next suffix. What is perhaps more interesting is that the $C_1$ and $C_2$ codes of Apostolico and Fraenkel are special cases of this code. Thus, $C_1^{(3)}$ is obtained from $S(4,0111)$ by replacing all the sequences that start with 11 with the sequence 11 and moving the rightmost 1 to the left end of the codeword. Also, $C_2^{(2)}$ is obtained from $S(3,011)$ by replacing all the sequences that start with 1 with the codeword 1 and moving the rightmost 1 to the left end of the codeword.

The developer provides algorithms for encoding and decoding this code.

> Fibonacci couldn't sleep—
> Counted rabbits instead of sheep.
> —Katherine O'Brien

# 2.21 Goldbach Codes

The Fibonacci codes of Section 2.19 have the rare property, termed Zeckendorf's theorem, that they do not have consecutive 1's. If a binary number of the form $1xx\ldots x$ does not have any consecutive 1's, then reversing its bits and appending a 1 results in a number of the form $xx\ldots x1|1$ that ends with two consecutive 1's, thereby making it easy for a decoder to read individual, variable-length Fibonacci codewords from a long stream of such codes. The variable-length codes described in this section are based on a similar property that stems from the Goldbach conjecture.

> Goldbach's conjecture: Every even integer $n$ greater than 2 is the sum of two primes.

A prime number is an integer that is not divisible by any other integer (other than trivially by itself and by 1). The integer 2 is prime, but all other primes are odd. Adding two odd numbers results in an even number, so mathematicians have always known that the sum of two primes is even. History had to wait until 1742, when it occurred to Christian Goldbach, an obscure German mathematician, to ask the "opposite" question. If the sum of any two primes is even, is it true that every even integer is the sum of two primes? Goldbach was unable to prove this simple-looking problem, but neither was he able to find a counter-example. He wrote to the famous mathematician Leonhard Euler and received the answer "There is little doubt that this result is true." However, even Euler was unable to furnish a proof, and at the time of this writing (late 2006), after more than 260 years of research, the Goldbach conjecture has almost, but not quite, been proved. It should be noted that many even numbers can be written as the sum of two primes in several ways. Thus $42 = 23 + 19 = 29 + 13 = 31 + 11 = 37 + 5$, 1,000,000 can be partitioned in 5,402 ways, and 100,000,000 has 291,400 Goldbach partitions. Reference [pass 06] is an online calculator that can compute the Goldbach partitions of even integers.

Christian Goldbach was a Prussian mathematician. Born in 1690, the
son of a pastor, in Königsberg (East Prussia), Goldbach studied law
and mathematics. He traveled widely throughout Europe and met with
many well known mathematicians, such as Gottfried Leibniz, Leonhard
Euler, and Nicholas (I) Bernoulli. He went to work at the newly opened
St Petersburg Academy of Sciences and became tutor to the later Tsar
Peter II. The following quotation, from [Mahoney 90] reflects the feelings
of his superiors in Russia "... a superb command of Latin style and equal fluency in
German and French. Goldbach's polished manners and cosmopolitan circle of friends
and acquaintances assured his success in an elite society struggling to emulate its
western neighbors."
Goldbach is remembered today for Goldbach's conjecture. He also studied and proved
some theorems on perfect powers. He died in 1764.

In 2001, Peter Fenwick had the idea of using the Goldbach conjecture (assuming that
it is true) to design an entirely new class of codes, based on prime numbers [Fenwick 02].
The prime numbers can serve as the basis of a number system, so if we write an even
integer in this system, its representation will have exactly two 1's. Thus, the even
number 20 equals $7 + 13$ and can therefore be written 10100, where the five bits are
assigned the prime weights (from left to right) 13, 11, 7, 5, and 3. Now reverse this bit
pattern so that its least-significant bit becomes 1, to yield 00101. Such a number is easy
to read and extract from a long bitstring. Simply stop reading at the second 1. Recall
that the unary code (a sequence of zeros terminated by a single 1) is read by a similar
rule: stop at the first 1. Thus, the Goldbach codes can be considered an extension of
the simple unary code.

| $n$ | $2(n+3)$ | Primes | Codeword |
|---|---|---|---|
| 1 | 8 | $3+5$ | 11 |
| 2 | 10 | $3+7$ | 101 |
| 3 | 12 | $5+7$ | 011 |
| 4 | 14 | $3+11$ | 1001 |
| 5 | 16 | $5+11$ | 0101 |
| 6 | 18 | $7+11$ | 0011 |
| 7 | 18 | $7+13$ | 00101 |
| 8 | 22 | $5+17$ | 010001 |
| 9 | 24 | $11+13$ | 00011 |
| 10 | 26 | $7+19$ | 0010001 |
| 11 | 28 | $11+17$ | 000101 |
| 12 | 30 | $13+17$ | 000011 |
| 13 | 32 | $13+19$ | 0000101 |
| 14 | 34 | $11+23$ | 00010001 |

Table 2.38: The Goldbach G0 Code.

The first Goldbach code is designated G0. It encodes the positive integer $n$ by
examining the even number $2(n+3)$ and writing it as the sum of two primes in reverse
(with its most-significant bit at the right end). The G0 codes listed in Table 2.38 are

based on the primes (from right to left) 23, 19, 17, 13, 11, 7, 5, and 3. It is obvious that the codes increase in length, but not monotonically, and there is no known expression for the length of G0($n$) as a function of $n$

---

Goldbach's original conjecture (sometimes called the "ternary" Goldbach conjecture), written in a June 7, 1742 letter to Euler [dartmouth 06], states "at least it seems that every integer that is greater than 2 is the sum of three primes." This made sense because Goldbach considered 1 a prime, a convention that is no longer followed. Today, his statement would be rephrased to "every integer greater than 5 is the sum of three primes." Euler responded with "There is little doubt that this result is true," and coined the modern form of the conjecture (a form currently referred to as the "strong" or "binary" Goldbach conjecture) which states "all positive even integers greater than 2 can be expressed as the sum of two primes." Being honest, Euler also added in his response that he regarded this as a fully certain theorem ("ein ganz gewisses Theorema"), even though he was unable to prove it.

Reference [utm 06] has information on the history of this and other interesting mathematical conjectures.

---

The G0 codes are efficient for small values of $n$ because (1) they are easy to construct based on a table of primes, (2) they are easy to decode, and (3) they are about as long as the Fibonacci codes or the standard binary representation (the $\beta$ code) of the integers. However, for large values of $n$, the G0 codes become too long because, as Table 2.39 illustrates, the primes are denser than the powers of 2 or the Fibonacci numbers. Also, a large even number can normally be written as the sum of two primes in many different ways. For large values of $n$, a large table of primes is therefore needed and it may take the encoder a while to determine the pair of primes that yields the shortest code for a given large integer $n$. (The shortest code is normally produced by two primes of similar sizes. Writing $n = a + b$ where $a$ is small and $b$ is large, results in a long G0 code. The best Goldbach partition for 11,230, for example, is $2003 + 9227$ and the worst one is $17 + 11213$.)

| $n$: | 1 | 2 | 3 | 4 | 5 | 6 | 7 | 8 | 9 | 10 | 11 | 12 |
|------|---|---|---|---|---|---|---|---|---|----|----|----|
| $P_n$: | 1 | 3 | 5 | 7 | 11 | 13 | 17 | 19 | 23 | 29 | 31 | 37 |
| $2^{n-1}$: | 1 | 2 | 4 | 8 | 16 | 32 | 64 | 128 | 256 | 512 | 1024 | 2048 |
| $F_n$: | 1 | 1 | 2 | 3 | 5 | 8 | 13 | 21 | 34 | 55 | 89 | 144 |

Table 2.39: Growth of Primes, Powers of 2, and Fibonacci Numbers.

Thus, the G0 code of a large integer $n$ is long, and since it has only two 1's, it must have many zeros and may have one or two runs of zeros. This property is the basis for the Goldbach G1 code. The principle of G1 is to determine two primes $P_i$ and $P_j$ (where $i \leq j$) whose sum yields a given integer $n$, and encode the pair $(i, j - i + 1)$ with two gamma codes. Thus, for example, $n = 100$ can be written as the following sums $3 + 97$, $11 + 89$, $17 + 83$, $29 + 71$, $41 + 59$, and $47 + 53$. We select $47 + 53 = P_{15} + P_{16}$, yielding the pair $(15, 16 - 15 + 1 = 2)$ and the two gamma codes 0001111:010 that are concatenated to form the G1 code of 100. For comparison, selecting the Goldbach partition $100 = 3 + 97$ yields the indexes 2 and 25, the pair $(2, 25 - 2 + 1)$, and the two gamma codes 010:000011000, two bits longer. Notice that $i$ may equal $j$, which is why the pair $(i, j - i + 1)$ and not $(i, j - i)$ is encoded. The latter may result in a second element of 0.

Table 2.40 lists several G1 codes of even integers. It is again obvious that the lengths of the G1 codes increase but not monotonically. The lengths of the corresponding gamma codes are also listed for comparison, and it is clear that the G1 code is the winner in most cases. Figure 2.41 illustrates the lengths of the G1, Elias gamma code, and the $C^1$ code of Fraenkel and Klein (Section 2.19).

| $n$ | Sum | Indexes | Pair | Codeword | Len. | $|\gamma(n)|$ |
|-----|------|---------|------|----------|------|---------------|
| 2 | 1+1 | 1,1 | 1,1 | 1:1 | 2 | 3 |
| 4 | 1+3 | 1,2 | 1,2 | 1:010 | 4 | 5 |
| 6 | 3+3 | 2,2 | 2,1 | 010:1 | 4 | 5 |
| 8 | 3+5 | 2,3 | 2,2 | 010:010 | 6 | 7 |
| 10 | 3+7 | 2,4 | 2,3 | 010:011 | 6 | 7 |
| 12 | 5+7 | 3,4 | 3,2 | 011:010 | 6 | 7 |
| 14 | 7+7 | 4,4 | 4,1 | 00100:1 | 6 | 7 |
| 16 | 5+11 | 3,5 | 3,3 | 011:011 | 6 | 9 |
| 18 | 7+11 | 4,5 | 4,2 | 00100:011 | 8 | 9 |
| 20 | 7+13 | 4,6 | 4,3 | 00100:011 | 8 | 9 |
| 30 | 13+17 | 6,7 | 6,2 | 00110:010 | 8 | 9 |
| 40 | 17+23 | 7,9 | 7,3 | 00111:011 | 8 | 11 |
| 50 | 13+37 | 6,12 | 6,7 | 00110:00111 | 10 | 11 |
| 60 | 29+31 | 10,11 | 10,2 | 0001010:010 | 10 | 11 |
| 70 | 29+41 | 10,13 | 10,4 | 0001010:00100 | 12 | 13 |
| 80 | 37+43 | 12,14 | 12,3 | 0001100:011 | 10 | 13 |
| 90 | 43+47 | 14,15 | 14,2 | 0001110:010 | 10 | 13 |
| 100 | 47+53 | 15,16 | 15,2 | 0001111:010 | 10 | 13 |
| 40 | 3+37 | 2,12 | 2,11 | 010:0001011 | 10 | 11 |
| 40 | 11+29 | 5,10 | 5,6 | 00101:00110 | 10 | 11 |

Table 2.40: The Goldbach G1 Code.

The last two rows of Table 2.40 list two alternatives for the G1 code of 40, and they really are two bits longer than the best code of 40. However, selecting the Goldbach partition with the most similar primes does not always yield the shortest code, as in the case of $50 = 13 + 37 = 19 + 31$. Selecting the first pair (where the two primes differ most) yields the indexes $6, 12$ and the pair $(6, 7)$ for a total gamma codes of 10 bits. Selecting the second pair (the most similar primes), on the other hand, results in indexes $8, 11$ and a pair $(8, 4)$ for a total of 12 bits of gamma codes.

In order for it to be useful, the G1 code has to be extended to arbitrary positive integers, a task that is done in two ways as follows:

1. Map the positive integer $n$ to the even number $N = 2(n + 3)$ and encode $N$ in G1. This is a natural extension of G1, but it results in indexes that are about 60% larger and thus generate long codes. For example, the extended code for 20 would be

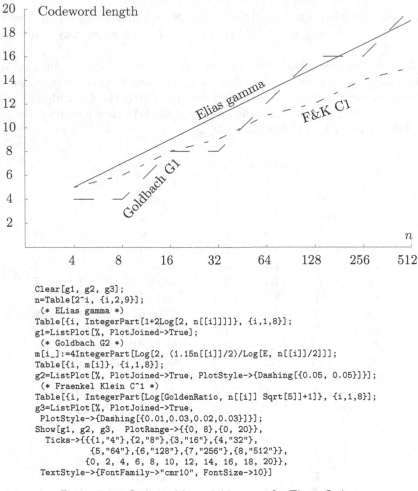

```
Clear[g1, g2, g3];
n=Table[2^i, {i,2,9}];
 (* ELias gamma *)
Table[{i, IntegerPart[1+2Log[2, n[[i]]]]}, {i,1,8}];
g1=ListPlot[%, PlotJoined->True];
 (* Goldbach G2 *)
m[i_]:=4IntegerPart[Log[2, (1.15n[[i]]/2)/Log[E, n[[i]]/2]]];
Table[{i, m[i]}, {i,1,8}];
g2=ListPlot[%, PlotJoined->True, PlotStyle->{Dashing[{0.05, 0.05}]}];
 (* Fraenkel Klein C^1 *)
Table[{i, IntegerPart[Log[GoldenRatio, n[[i]] Sqrt[5]]+1]}, {i,1,8}];
g3=ListPlot[%, PlotJoined->True,
 PlotStyle->{Dashing[{0.01,0.03,0.02,0.03}]}];
Show[g1, g2, g3,  PlotRange->{{0, 8},{0, 20}},
  Ticks->{{{1,"4"},{2,"8"},{3,"16"},{4,"32"},
          {5,"64"},{6,"128"},{7,"256"},{8,"512"}},
          {0, 2, 4, 6, 8, 10, 12, 14, 16, 18, 20}},
 TextStyle->{FontFamily->"cmr10", FontSize->10}]
```

Figure 2.41: Codeword Length Versus $n$ for Three Codes.

the original G1 code of $46 = 17 + 29$. The indexes are $7, 10$, the pair is $(7, 4)$, and the codes are 00111:00100, two bits longer than the original G1 code of 20.

2. A better way to extend G1 is to consider various cases and handle each differently so as to obtain the best codes in every case. The developer, Peter Fenwick, refers to the result as the G2 code. The cases are as follows:

2.1. The integers 1 and 2 are encoded as 110 and 111, respectively (no other codes will start with 11).

2.2. The even integers are encoded as in G1, but with a small difference. Once it is determined that $n = P_i + P_j$, we encode the pair $(i + 1, j - i + 1)$ instead of $(i, j - i + 1)$. Thus, if $i = 1$, it is encoded as 010, the gamma code of 2. This guarantees that the G2 code of an even integer will not start with a 1 and will always have the form $0\ldots:0\ldots$.

2.3. If $n$ is the prime $P_i$, it is encoded as the gamma code of $(i + 1)$ followed by a single 1 to yield $0\ldots:1$.

2.4. If $n$ is odd but is not a prime, its G2 code starts with a single 1 followed

by the G2 code of the even number $n - 1$. The resulting gamma codes have the form
$1:0\ldots:0\ldots$

Table 2.42 lists some examples of the G2 code and compares their lengths to the
lengths of the gamma and omega codes. In most cases, G2 is the shortest of the three,
but in some cases, most notably when $n$ is a power of 2, G2 is longer than the other
codes.

| $n$ | Codeword | Len. | $|\gamma(n)|$ | $|\omega(n)|$ |
|---|---|---|---|---|
| 1 | 110 | 3 | 1 | 1 |
| 2 | 111 | 3 | 3 | 3 |
| 3 | 0101 | 4 | 3 | 3 |
| 4 | 010010 | 6 | 5 | 6 |
| 5 | 0111 | 4 | 5 | 6 |
| 6 | 011010 | 6 | 5 | 6 |
| 7 | 001001 | 6 | 5 | 6 |
| 8 | 011011 | 6 | 7 | 7 |
| 9 | 1011011 | 7 | 7 | 7 |
| 10 | 01100100 | 8 | 7 | 7 |
| 11 | 001011 | 6 | 7 | 7 |
| 12 | 00101010 | 8 | 7 | 7 |
| 13 | 001101 | 6 | 7 | 7 |
| 14 | 00110010 | 8 | 7 | 7 |
| 15 | 100110010 | 9 | 7 | 7 |
| 16 | 0010100100 | 10 | 9 | 11 |
| 17 | 001111 | 6 | 9 | 11 |
| 18 | 00111010 | 8 | 9 | 11 |
| 19 | 00010001 | 8 | 9 | 11 |
| 20 | 00111011 | 8 | 9 | 11 |
| 30 | 0001010010 | 10 | 9 | 11 |
| 40 | 0001100011 | 10 | 11 | 12 |
| 50 | 0001111011 | 10 | 11 | 12 |
| 60 | 000010001010 | 12 | 11 | 12 |
| 70 | 000010011011 | 12 | 13 | 13 |
| 80 | 000010110010 | 12 | 13 | 13 |
| 90 | 000011000010 | 12 | 13 | 13 |
| 100 | 000011001011 | 12 | 13 | 13 |

Table 2.42: The Goldbach G2 Code.

Given an even integer $n$, there is no known formula to locate any of its Goldbach
partitions. Thus, we cannot determine the precise length of any Goldbach code, but we
can provide an estimate, and the G1 code length is the easiest to estimate. We assume
that the even number $2n$ is the sum of two primes, each of order $n$. According to the
prime number theorem, the number $\pi(n)$ of primes less than $n$ is approximately proportional to $n/\ln n$. Extensive experiments carried out by Peter Fenwick, the developer of
these codes, indicate that for values up to 1000, a good constant of proportionality is

1.15. Thus, we estimate index $i$ at $i \approx 1.15n/\ln n$, resulting in a gamma code of

$$2 \left\lfloor \log_2 \frac{1.15n}{\ln n} \right\rfloor + 1 \stackrel{\text{def}}{=} 2L + 1 \quad \text{bits.} \tag{2.10}$$

We estimate the second element of the pair $(i, j - i + 1)$ at $i/2$ (Table 2.40 shows that it is normally smaller), which implies that the second gamma code is two bits shorter than the first. The total length of the G1 code is therefore $(2L + 1) + (2L - 1) = 4L$, where $L$ is given by Equation (2.10). Direct computations show that the G1 code is generally shorter than the gamma code for values up to 100 and longer for larger values.

The length of the G2 code is more difficult to estimate, but a simple experiment that computed these codes for $n$ values from 2 to 512 indicates that their lengths (which often vary by 2–3 bits from code to code) can be approximated by the smooth function $2 + \frac{13}{8} \log_2 n$. For $n = 512$, this expression has the value 16.625, which is 1.66 times the length of the binary representation of 512 (10 bits). This should be contrasted with the gamma code, where the corresponding factor is 2.

> I can envision an abstract of a paper, circa 2100, that reads: "We can show, in a certain precise sense, that the Goldbach conjecture is true with probability larger than 0.99999, and that its complete truth could be determined with a budget of \$10B.")
> —Doron Zeilberger (1993)

## 2.22 Additive Codes

The fact that the Fibonacci numbers and the Goldbach conjecture lead to simple, efficient codes suggests that other number sequences may be employed in the same way to create similar, and perhaps even better, codes. Specifically, we may extend the Goldbach codes if we find a sequence $S$ of "basis" integers such that any integer $n$ can be expressed as the sum of two elements of $S$ (it is possible to search for similar sequences that can express any $n$ as the sum of three, four, or more elements, but we restrict our search to sums of two sequence elements). Given such a sequence, we can conceive a code similar to G0, but for all the nonnegative integers, not just the even integers. Given several such sequences, we can call the resulting codes "additive codes."

One technique to generate a sequence of basis integers is based on the sieve principle (compare with the sieve of Eratosthenes). We start with a short sequence (a basis set) whose elements $a_i$ are sufficient to represent each of the first few positive integers as a sum $a_i + a_j$. This initial sequence is then enlarged in steps. In a general step, we have just added a new element $a_k$ and ended up with the sequence $S = (0, a_1, a_2, \ldots, a_k)$ such that each integer $1 \leq n \leq a_k$ can be represented as a sum of two elements of $S$. We now generate all the sums $a_i + a_k$ for $i$ values from 1 to $k$ and check to verify that they include $a_k + 1$, $a_k + 2$, and so on. When we find the first missing integer, we append it to $S$ as element $a_{k+1}$. We now know that each integer $1 \leq n \leq a_{k+1}$ can be represented as a sum of two elements from $S$, and we can continue to extend sequence $S$ of basis integers.

This sieve technique is illustrated with the basis set $(0, 1, 2)$. Any of 0, 1, and 2 can be represented as the sum of two elements from this set. We generate all the possible sums $a_i + a_k = a_i + 2$ and obtain 3 and 4. Thus, 5 is the next integer that cannot be generated, and it becomes the next element of the set. Adding $a_i = 5$ yields 5, 6, and 7, so the next element is 8. Adding $a_i + 8$ in the sequence $(0, 1, 2, 5, 8)$ yields 8, 9, 10, 13, and 16, so the next element should be 11. Adding $a_i + 11$ in the sequence $(0, 1, 2, 5, 8, 11)$ yields 11, 12, 13, 16, 19, and 22, so the next element should be 14. Continuing in this way, we obtain the sequence 0, 1, 2, 5, 8, 11, 14, 16, 20, 23, 26, 29, 33, 46, 50, 63, 67, 80, 84, 97, 101, 114, 118, 131, 135, 148, 152, 165, 169, 182, 186, 199, 203, 216, 220, 233, and 237 whose 37 elements can represent every integer from 0 to 250 as the sum of two elements. The equivalent Goldbach code for integers up to 250 requires 53 primes, so it has the potential of being longer.

The initial basis set may be as small as $(0, 1)$, but may also contain other integers (seeds). In fact, the seeds determine the content and performance of the additive sequence. Starting with just the smallest basic set $(0, 1)$, adding $a_i + 1$ yields 1 and 2, so the next element should be 3. Adding $a_i + 3$ in the sequence $(0, 1, 3)$ yields 3, 4, and 6, so the next element should be 5. Adding $a_i + 5$ in the sequence $(0, 1, 3, 5)$ yields 5, 6, 8, and 10, so the next element should be 7. We end up with a sequence of only odd numbers, which is why it may be a good idea to start with the basic sequence plus some even seeds. Table 2.43 lists some additive codes based on the additive sequence $(0, 1, 3, 5, 7, 9, 11, 12, 25, 27, 29, 31, 33, 35, \ldots)$ and compares their lengths to the lengths of the corresponding gamma codes.

| n | Sum | Indexes | Pair | Codeword | Len. | $|\gamma(n)|$ |
|---|---|---|---|---|---|---|
| 10 | $3 + 7$ | 3,5 | 3,3 | 011:011 | 6 | 7 |
| 11 | $0 + 11$ | 1,7 | 1,7 | 1:00111 | 6 | 7 |
| 12 | $1 + 11$ | 2,7 | 2,6 | 010:00110 | 8 | 7 |
| 13 | $1 + 12$ | 2,8 | 2,7 | 010:00111 | 8 | 7 |
| 14 | $7 + 7$ | 5,5 | 5,1 | 00101:1 | 6 | 7 |
| 15 | $3 + 12$ | 3,8 | 3,6 | 011:00110 | 8 | 7 |
| 16 | $7 + 9$ | 5,6 | 5,2 | 00101:010 | 8 | 9 |
| 17 | $5 + 12$ | 4,8 | 4,5 | 00100:00101 | 10 | 9 |
| 18 | $7 + 11$ | 5,7 | 5,3 | 00101:011 | 8 | 9 |
| 20 | $9 + 11$ | 6,7 | 6,2 | 00110:010 | 8 | 9 |
| 30 | $5 + 25$ | 4,9 | 4,6 | 00100:00110 | 10 | 9 |
| 40 | $11 + 29$ | 7,11 | 7,5 | 00111:00101 | 10 | 11 |
| 50 | $25 + 25$ | 9,9 | 9,1 | 0001001:1 | 8 | 11 |
| 60 | $29 + 31$ | 11,12 | 11,2 | 0001011:010 | 10 | 11 |
| 70 | $35 + 35$ | 14,14 | 14,1 | 0001110:1 | 8 | 13 |
| 80 | $0 + 80$ | 1,20 | 1,20 | 1:000010100 | 10 | 13 |
| 90 | $29 + 61$ | 11,17 | 11,7 | 0001011:00111 | 14 | 13 |
| 100 | $3 + 97$ | 3,23 | 3,21 | 011:000010101 | 14 | 13 |

Table 2.43: An Additive Code.

The developer of these codes presents a summary where the gamma, delta,

Fibonacci, G1, and additive codes are compared for integers $n$ from 1 to 256, and concludes that for values from 4 to 30, the additive code is the best of the five, and for the entire range, it is better than the gamma and delta codes.

Given integer data in a certain interval, it seems that the best additive codes to compress the data are those generated by the shortest additive sequence. Determining this sequence can be done by a brute force approach that tries many sets of seeds and computes the additive sequence generated by each set.

---

### Stanislaw Marcin Ulam (1909–1984)

One of the most prolific mathematicians of the 20th century, Stanislaw Ulam also had interests in astronomy and physics. He worked on the hydrogen bomb at the Los Alamos National Laboratory, proposed the Orion project for nuclear propulsion of space vehicles, and originated the Monte-Carlo method (and also coined this term). In addition to his important original work, mostly in point set topology, Ulam was also interested in mathematical recreations, games, and oddities. The following quotation, by his friend and colleague Gian-Carlo Rota, summarizes Ulam's unusual personality and talents.

"Ulam's mind is a repository of thousands of stories, tales, jokes, epigrams, remarks, puzzles, tongue-twisters, footnotes, conclusions, slogans, formulas, diagrams, quotations, limericks, summaries, quips, epitaphs, and headlines. In the course of a normal conversation he simply pulls out of his mind the fifty-odd relevant items, and presents them in linear succession. A second-order memory prevents him from repeating himself too often before the same public."

There is another Ulam sequence that is constructed by a simple rule. Start with any positive integer $n$ and construct a sequence as follows:

1. If $n = 1$, stop.
2. If $n$ is even, the next number is $n/2$; go to step 1.
3. If $n$ is odd, the next number is $3n + 1$; go to step 1.

Here are some examples for the first few integers: (1), (10, 5, 16, 8, 4, 2, 1), (2, 1), (16, 8, 4, 2, 1), (3, 10, 5, ...) (22, 11, 34, 17, 52, 26, 13, 40, 20, 10, 5, ...).

These sequences are sometimes known as the $3x + 1$ problem, because no one has proved that they always reach the value 1. However, direct checking up to $100 \times 2^{50}$ suggests that this may be so.

---

**Ulam Sequence.** The Ulam sequence $(u, v)$ is defined by $a_1 = u$, $a_2 = v$, and $a_i$ for $i > 2$ is the smallest integer that can be expressed *uniquely* as the sum of two distinct earlier elements. The numbers generated this way are sometimes called u-numbers or Ulam numbers. A basic reference is [Ulam 06].

The first few elements of the $(1, 2)$-Ulam sequence are 1, 2, 3, 4, 6, 8, 11, 13, 16,.... The element following the initial $(1, 2)$ is 3, because $3 = 1 + 2$. The next element is $4 = 1 + 3$. (There is no need to worry about $4 = 2 + 2$, because this is a sum of two identical elements instead of two distinct elements.) The integer 5 is not an element of the sequence because it is not uniquely representable, but $6 = 2 + 4$ is.

These simple rules make it possible to generate Ulam sequences for any pair $(u, v)$ of integers. Table 2.44 lists several example (the "Sloane" labels in this example column refer to integer sequences from [Sloane 06]).

| $(u, v)$ | Sloane | Sequence |
|----------|---------|----------|
| $(1, 2)$ | A002858 | 1, 2, 3, 4, 6, 8, 11, 13, 16, 18, ... |
| $(1, 3)$ | A002859 | 1, 3, 4, 5, 6, 8, 10, 12, 17, 21, ... |
| $(1, 4)$ | A003666 | 1, 4, 5, 6, 7, 8, 10, 16, 18, 19, ... |
| $(1, 5)$ | A003667 | 1, 5, 6, 7, 8, 9, 10, 12, 20, 22, ... |
| $(2, 3)$ | A001857 | 2, 3, 5, 7, 8, 9, 13, 14, 18, 19, ... |
| $(2, 4)$ | A048951 | 2, 4, 6, 8, 12, 16, 22, 26, 32, 36, ... |
| $(2, 5)$ | A007300 | 2, 5, 7, 9, 11, 12, 13, 15, 19, 23, ... |

Table 2.44: Some Ulam Sequences.

It is clear that Ulam sequences are additive and can be used to generate additive codes. The following *Mathematica* code to generate such a sequence is from [Sloane 06], sequence A002858.

```
Ulam4Compiled = Compile[{{nmax, _Integer}, {init, _Integer, 1}, {s, _Integer}},
  Module[{ulamhash = Table[0, {nmax}], ulam = init},
  ulamhash[[ulam]] = 1;
  Do[ If[Quotient[Plus @@ ulamhash[[i - ulam]], 2] == s, AppendTo[ulam, i];
  ulamhash[[i]] = 1], {i, Last[init] + 1, nmax}]; ulam]];
Ulam4Compiled[355, {1, 2}, 1]
```

# 2.23 Golomb Code

The seventeenth-century French mathematician Blaise Pascal is known today mostly for his contributions to the field of probability, but during his short life he made important contributions to many areas. It is generally agreed today that he invented (an early version of) the game of roulette (although some believe that this game originated in China and was brought to Europe by Dominican monks who were trading with the Chinese). The modern version of roulette appeared in 1842.

The roulette wheel has 37 shallow depressions (known as slots) numbered 0 through 36 (the American version has 38 slots numbered 00, 0, and 1 through 36). The dealer (croupier) spins the wheel while sending a small ball rolling in the opposite direction inside the wheel. Players can place bets during the spin until the dealer says "no more bets." When the wheel stops, the slot where the ball landed determines the outcome of the game. Players who bet on the winning number are paid according to the type of bet they placed, while players who bet on the other numbers lose their entire bets to the house. [Bass 92] is an entertaining account of an attempt to compute the result of a roulette spin in real time.

The simplest type of bet is on a single number. A player winning this bet is paid 35 times the amount bet. Thus, a player who plays the game repeatedly and bets \$1 each time expects to lose 36 games and win one game out of every set of 37 games on average. The player therefore loses on average \$37 for every \$35 won.

The probability of winning a game is $p = 1/37 \approx 0.027027$ and that of losing a game is the much higher $q = 1 - p = 36/37 \approx 0.972973$. The probability $P(n)$ of winning once and losing $n - 1$ times in a sequence of $n$ games is the product $q^{n-1}p$. This probability is normalized because

$$\sum_{n=1}^{\infty} P(n) = \sum_{n=1}^{\infty} q^{n-1}p = p\sum_{n=0}^{\infty} q^n = \frac{p}{1-q} = \frac{p}{p} = 1.$$

As $n$ grows, $P(n)$ shrinks slowly because of the much higher value of $q$. The values of $P(n)$ for $n = 1, 2, \ldots, 10$ are 0.027027, 0.026297, 0.025586, 0.024895, 0.024222, 0.023567, 0.022930, 0.022310, 0.021707, and 0.021120.

The probability function $P(n)$ is said to obey a *geometric distribution*. The reason for the name "geometric" is the resemblance of this distribution to the geometric sequence. A sequence where the ratio between consecutive elements is a constant $q$ is called geometric. Such a sequence has elements $a$, $aq$, $aq^2$, $aq^3$, .... The (infinite) sum of these elements is a geometric series $\sum_{i=0}^{\infty} aq^i$. The interesting case is where $q$ satisfies $-1 < q < 1$, in which the series converges to $a/(1-q)$. Figure 2.45 shows the geometric distribution for $p = 0.2$, 0.5, and 0.8.

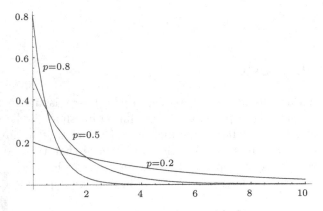

Figure 2.45: Geometric Distributions for $p = 0.2$, 0.5, and 0.8.

Certain data compression methods are based on run-length encoding (RLE). Imagine a binary string where a 0 appears with probability $p$ and a 1 appears with probability $1 - p$. If $p$ is large, there will be runs of zeros, suggesting the use of RLE to compress the string. The probability of a run of $n$ zeros is $p^n$, and the probability of a run of $n$ zeros followed by a 1 is $p^n(1 - p)$, indicating that run lengths are distributed geometrically. A naive approach to compressing such a string is to compute the probability of each run length and apply the Huffman method to obtain the best prefix codes for the run lengths. In practice, however, there may be a large number of run lengths and this number may not be known in advance. A better approach is to construct an infinite family of individual prefix codes, such that no matter how long a run is, there will be a code in the family to encode it. The codes in the family must depend on the probability $p$, so we are looking for an infinite family of *parametrized* prefix codes. The Golomb codes presented here [Golomb 66] are such codes and they are the best ones for the compression of data items that are distributed geometrically.

Let's first examine a few numbers to see why such codes must depend on $p$. For $p = 0.99$, the probabilities of runs of two zeros and of 10 zeros are $0.99^2 = 0.9801$ and $0.99^{10} = 0.9$, respectively (both large). In contrast, for $p = 0.6$, the same run lengths have the much smaller probabilities of 0.36 and 0.006. The ratio $0.9801/0.36$ is 2.7225, but the ratio $0.9/0.006$ is the much greater 150. Thus, a large $p$ implies higher probabilities for long runs, whereas a small $p$ implies that long runs will be rare.

Two relevant statistical concepts are the mean and median of a sequence of run lengths. They are illustrated by the binary string

$$00000100110001010000001110100010000010001001000110100001001 \qquad (2.11)$$

that has the 18 run lengths 5, 2, 0, 3, 1, 6, 0, 0, 1, 3, 5, 3, 2, 3, 0, 1, 4, and 2. Its mean is the average $(5+2+0+3+1+6+0+0+1+3+5+3+2+3+0+1+4+2)/18 \approx 2.28$. Its median $m$ is the value such that about half the run lengths are shorter than $m$ and about half are equal to or greater than $m$. To find $m$, we sort the 18 run lengths to obtain 0, 0, 0, 0, 1, 1, 1, 2, 2, 2, 3, 3, 3, 3, 4, 5, 5, and 6 and find that the median (the central number) is 2.

We are now ready for a description of the Golomb code. The main feature of this code is its coding efficiency when the data consists of two asymmetric events, one common and the other one rare, that are interleaved.

**Encoding.** The Golomb code for nonnegative integers $n$ depends on the choice of a parameter $m$ (we'll see later that for RLE, $m$ should depend on the probability $p$ and on the median of the run lengths). Thus, it is a parametrized prefix code, which makes it especially useful in cases where good values for the parameter can be computed or estimated. The first step in constructing the Golomb code of the nonnegative integer $n$ is to compute the three quantities $q$ (quotient), $r$ (remainder), and $c$ by

$$q = \left\lfloor \frac{n}{m} \right\rfloor, \quad r = n - qm, \text{ and } c = \lceil \log_2 m \rceil,$$

following which the code is constructed in two parts; the first is the value of $q$, coded in unary, and the second is the binary value of $r$ coded in a special way. The first $2^c - m$ values of $r$ are coded, as unsigned integers, in $c - 1$ bits each, and the rest are coded in $c$ bits each (ending with the biggest $c$-bit number, which consists of $c$ 1's). The case where $m$ is a power of 2 ($m = 2^c$) is special because it requires no $(c-1)$-bit codes. We know that $n = r + qm$; so once a Golomb code is decoded, the values of $q$ and $r$ can be used to easily reconstruct $n$. The case $m = 1$ is also special. In this case, $q = n$ and $r = c = 0$, implying that the Golomb code of $n$ is its unary code.

**Examples.** Choosing $m = 3$ produces $c = 2$ and the three remainders 0, 1, and 2. We compute $2^2 - 3 = 1$, so the first remainder is coded in $c - 1 = 1$ bit to become 0, and the remaining two are coded in two bits each ending with $11_2$, to become 10 and 11. Selecting $m = 5$ results in $c = 3$ and produces the five remainders 0 through 4. The first three ($2^3 - 5 = 3$) are coded in $c - 1 = 2$ bits each, and the remaining two are each coded in three bits ending with $111_2$. Thus, 00, 01, 10, 110, and 111. The following simple rule shows how to encode the $c$-bit numbers such that the last of them will consist of $c$ 1's. Denote the largest of the $(c - 1)$-bit numbers by $b$, then construct the integer $b + 1$ in $c - 1$ bits, and append a zero on the right. The result is the first of the $c$-bit numbers and the remaining ones are obtained by incrementing.

Table 2.46 shows some examples of $m$, $c$, and $2^c - m$, as well as some Golomb codes for $m = 2$ through 13.

| $m$ | 2 | 3 | 4 | 5 | 6 | 7 | 8 | 9 | 10 | 11 | 12 | 13 | 14 | 15 | 16 |
|---|---|---|---|---|---|---|---|---|---|---|---|---|---|---|---|
| $c$ | 1 | 2 | 2 | 3 | 3 | 3 | 3 | 4 | 4 | 4 | 4 | 4 | 4 | 4 | 4 |
| $2^c - m$ | 0 | 1 | 0 | 3 | 2 | 1 | 0 | 7 | 6 | 5 | 4 | 3 | 2 | 1 | 0 |

| $m/n$ | 0 | 1 | 2 | 3 | 4 | 5 | 6 | 7 | 8 | 9 | 10 | 11 | 12 |
|---|---|---|---|---|---|---|---|---|---|---|---|---|---|
| 2 | 0\|0 | 0\|1 | 10\|0 | 10\|1 | 110\|0 | 110\|1 | 1110\|0 | 1110\|1 | 11110\|0 | 11110\|1 | 111110\|0 | 111110\|1 | 1111110\|0 |
| 3 | 0\|0 | 0\|10 | 0\|11 | 10\|0 | 10\|10 | 10\|11 | 110\|0 | 110\|10 | 110\|11 | 1110\|0 | 1110\|10 | 1110\|11 | 11110\|0 |
| 4 | 0\|00 | 0\|01 | 0\|10 | 0\|11 | 10\|00 | 10\|01 | 10\|10 | 10\|11 | 110\|00 | 110\|01 | 110\|10 | 110\|11 | 11110\|00 |
| 5 | 0\|00 | 0\|01 | 0\|10 | 0\|110 | 0\|111 | 10\|00 | 10\|01 | 10\|10 | 10\|110 | 10\|111 | 110\|00 | 110\|01 | 110\|10 |
| 6 | 0\|00 | 0\|01 | 0\|100 | 0\|101 | 0\|110 | 0\|111 | 10\|00 | 10\|01 | 10\|100 | 10\|101 | 10\|110 | 10\|111 | 110\|00 |
| 7 | 0\|00 | 0\|010 | 0\|011 | 0\|100 | 0\|101 | 0\|110 | 0\|111 | 10\|00 | 10\|010 | 10\|011 | 10\|100 | 10\|101 | 10\|110 |
| 8 | 0\|000 | 0\|001 | 0\|010 | 0\|011 | 0\|100 | 0\|101 | 0\|110 | 0\|111 | 10\|000 | 10\|001 | 10\|010 | 10\|011 | 10\|100 |
| 9 | 0\|000 | 0\|001 | 0\|010 | 0\|011 | 0\|100 | 0\|101 | 0\|110 | 0\|1110 | 0\|1111 | 10\|000 | 10\|001 | 10\|010 | 10\|011 |
| 10 | 0\|000 | 0\|001 | 0\|010 | 0\|011 | 0\|100 | 0\|101 | 0\|1100 | 0\|1101 | 0\|1110 | 0\|1111 | 10\|000 | 10\|001 | 10\|010 |
| 11 | 0\|000 | 0\|001 | 0\|010 | 0\|011 | 0\|100 | 0\|1010 | 0\|1011 | 0\|1100 | 0\|1101 | 0\|1110 | 0\|1111 | 10\|000 | 10\|001 |
| 12 | 0\|000 | 0\|001 | 0\|010 | 0\|011 | 0\|1000 | 0\|1001 | 0\|1010 | 0\|1011 | 0\|1100 | 0\|1101 | 0\|1110 | 0\|1111 | 10\|000 |
| 13 | 0\|000 | 0\|001 | 0\|010 | 0\|0110 | 0\|0111 | 0\|1000 | 0\|1001 | 0\|1010 | 0\|1011 | 0\|1100 | 0\|1101 | 0\|1110 | 0\|1111 |

Table 2.46: Some Golomb Codes for $m = 2$ Through 13.

For a somewhat longer example, we select $m = 14$. This results in $c = 4$ and produces the 14 remainders 0 through 13. The first two ($2^4 - 14 = 2$) are coded in $c - 1 = 3$ bits each, and the remaining 12 are coded in four bits each, ending with $1111_2$ (and as a result starting with $0100_2$). Thus, we have 000, 001, followed by the 12 values 0100, 0101, 0110, 0111, ..., 1111. Table 2.47 lists several detailed examples and Table 2.48 lists 48 codes for $m = 14$ and for $m = 16$. The former starts with two 4-bit codes, followed by sets of 14 codes each that are getting longer by one bit. The latter is simpler because 16 is a power of 2. The Golomb codes for $m = 16$ consist of sets of 16 codes each that get longer by one bit. The Golomb codes for the case where $m$ is a power of 2 have been conceived by Robert F. Rice and are called Rice codes. They are employed by several algorithms for lossless audio compression.

Tables 2.47 and 2.48 illustrate the effect of $m$ on the code length. For small values of $m$, the Golomb codes start short and rapidly increase in length. They are appropriate for RLE in cases where the probability $p$ of a 0 bit is small, implying very few long runs. For large values of $m$, the initial codes (for $n = 1, 2, \ldots$) are long, but their lengths increase slowly. Such codes make sense for RLE when $p$ is large, implying that many long runs are expected.

| $n$ | 0 | 1 | 2 | 3 | ... | 13 | 14 | 15 | 16 | 17 | ... | 27 | 28 | 29 | 30 |
|---|---|---|---|---|---|---|---|---|---|---|---|---|---|---|---|
| $q = \lfloor \frac{n}{14} \rfloor$ | 0 | 0 | 0 | 0 | ... | 0 | 1 | 1 | 1 | 1 | ... | 1 | 2 | 2 | 2 |
| unary($q$) | 0 | 0 | 0 | 0 | ... | 0 | 10 | 10 | 10 | 10 | ... | 10 | 110 | 110 | 110 |
| $r$ | | | | | | | | | | | ... | | 000 | 001 | 0100 |

Table 2.47: Some Golomb Codes for $m = 14$.

| | $m = 14$ | | | | $m = 16$ | | | |
|---|---|---|---|---|---|---|---|---|
| $n$ | Code | $n$ | Code | | $n$ | Code | $n$ | Code |
| 0 | 0000 | 24 | 101100 | | 0 | 00000 | 24 | 101000 |
| 1 | 0001 | 25 | 101101 | | 1 | 00001 | 25 | 101001 |
| | | 26 | 101110 | | 2 | 00010 | 26 | 101010 |
| 2 | 00100 | 27 | 101111 | | 3 | 00011 | 27 | 101011 |
| 3 | 00101 | 28 | 110000 | | 4 | 00100 | 28 | 101100 |
| 4 | 00110 | 29 | 110001 | | 5 | 00101 | 29 | 101101 |
| 5 | 00111 | | | | 6 | 00110 | 30 | 101110 |
| 6 | 01000 | 30 | 1100100 | | 7 | 00111 | 31 | 101111 |
| 7 | 01001 | 31 | 1100101 | | 8 | 01000 | | |
| 8 | 01010 | 32 | 1100110 | | 9 | 01001 | 32 | 1100000 |
| 9 | 01011 | 33 | 1100111 | | 10 | 01010 | 33 | 1100001 |
| 10 | 01100 | 34 | 1101000 | | 11 | 01011 | 34 | 1100010 |
| 11 | 01101 | 35 | 1101001 | | 12 | 01100 | 35 | 1100011 |
| 12 | 01110 | 36 | 1101010 | | 13 | 01101 | 36 | 1100100 |
| 13 | 01111 | 37 | 1101011 | | 14 | 01110 | 37 | 1100101 |
| 14 | 10000 | 38 | 1101100 | | 15 | 01111 | 38 | 1100110 |
| 15 | 10001 | 39 | 1101101 | | | | 39 | 1100111 |
| | | 40 | 1101110 | | 16 | 100000 | 40 | 1101000 |
| 16 | 100100 | 41 | 1101111 | | 17 | 100001 | 41 | 1101001 |
| 17 | 100101 | 42 | 1110000 | | 18 | 100010 | 42 | 1101010 |
| 18 | 100110 | 43 | 1110001 | | 19 | 100011 | 43 | 1101011 |
| 19 | 100111 | | | | 20 | 100100 | 44 | 1101100 |
| 20 | 101000 | 44 | 11100100 | | 21 | 100101 | 45 | 1101101 |
| 21 | 101001 | 45 | 11100101 | | 22 | 100110 | 46 | 1101110 |
| 22 | 101010 | 46 | 11100110 | | 23 | 100111 | 47 | 1101111 |
| 23 | 101011 | 47 | 11100111 | | | | | |

Table 2.48: The First 48 Golomb Codes for $m = 14$ and $m = 16$.

**Decoding.** The Golomb codes are designed in this special way to facilitate their decoding. We first demonstrate the decoding for the simple case $m = 16$ ($m$ is a power of 2). To decode, start at the left end of the code and count the number $A$ of 1's preceding the first 0. The length of the code is $A + c + 1$ bits (for $m = 16$, this is $A + 5$ bits). If we denote the rightmost five bits of the code by $R$, then the value of the code is $16A + R$. This simple decoding reflects the way the code was constructed. To encode $n$ with $m = 16$, start by dividing it by 16 to get $n = 16A + R$, then write $A$ 1's followed by a single 0, followed by the 4-bit representation of $R$.

For $m$ values that are not powers of 2, decoding is slightly more involved. Assuming again that a code begins with $A$ 1's, start by removing them and the zero immediately following them. Denote the $c - 1$ bits that follow by $R$. If $R < 2^c - m$, then the total length of the code is $A + 1 + (c - 1)$ (the $A$ 1's, the zero following them, and the $c - 1$ bits that follow) and its value is $m \times A + R$. If $R \geq 2^c - m$, then the total length of the code is $A + 1 + c$ and its value is $m \times A + R' - (2^c - m)$, where $R'$ is the $c$-bit integer consisting of $R$ and the bit that follows $R$.

An example is the code $0001xxx$, for $m = 14$. There are no leading 1's, so $A$ is 0. After removing the leading zero, the $c - 1 = 3$ bits that follow are $R = 001$. Since $R < 2^c - m = 2$, we conclude that the length of the code is $0 + 1 + (4 - 1) = 4$ and its value is 001. Similarly, the code $00100xxx$ for the same $m = 14$ has $A = 0$ and $R = 010_2 = 2$. In this case, $R \geq 2^c - m = 2$, so the length of the code is $0 + 1 + c = 5$, the value of $R'$ is $0100_2 = 4$, and the value of the code is $14 \times 0 + 4 - 2 = 2$.

The JPEG-LS method for lossless image compression (recommendation ISO/IEC CD 14495) employs the Golomb code.

The family of Golomb codes has a close relative, the exponential Golomb codes which are described on page 164.

It is now clear that the best value for $m$ depends on $p$, and it can be shown that this value is the integer closest to $-1/\log_2 p$ or, equivalently, the value that satisfies

$$p^m \approx 1/2. \tag{2.12}$$

It can also be shown that in the case of a sequence of run lengths, this integer is the median of the run lengths. Thus, for $p = 0.5$, $m$ should be $-1/\log_2 0.5 = 1$. For $p = 0.7$, $m$ should be 2, because $-1/\log_2 0.7 \approx 1.94$, and for $p = 36/37$, $m$ should be 25, because $-1/\log_2(36/37) \approx 25.29$.

It should also be mentioned that Gallager and van Voorhis [Gallager and van Voorhis 75] have refined and extended Equation (2.12) into the more precise relation

$$p^m + p^{m+1} \leq 1 < p^m + p^{m-1}. \tag{2.13}$$

They proved that the Golomb code is the best prefix code when $m$ is selected by their inequality. We first show that for a given $p$, inequality (2.13) has only one solution $m$. We manipulate this inequality in four steps as follows:

$$p^m(1 + p) \leq 1 < p^{m-1}(1 + p),$$

$$p^m \leq \frac{1}{(1 + p)} < p^{m-1},$$

$$m \geq \frac{1}{\log p} \log \frac{1}{1 + p} > m - 1,$$

$$m \geq -\frac{\log(1 + p)}{\log p} > m - 1,$$

from which it is clear that the unique value of $m$ is

$$m = \left\lceil -\frac{\log_2(1 + p)}{\log_2 p} \right\rceil. \tag{2.14}$$

Three examples are presented here to illustrate the performance of the Golomb code in compressing run lengths. The first example is the binary string (2.11), which has 41 zeros and 18 ones. The probability of a zero is therefore $41/(41 + 18) \approx 0.7$, yielding $m = \lceil -\log 1.7/\log 0.7 \rceil = \lceil 1.487 \rceil = 2$. The sequence of run lengths 5, 2, 0, 3, 1, 6, 0, 0,

1, 3, 5, 3, 2, 3, 0, 1, 4, and 2 can therefore be encoded with the Golomb codes for $m = 2$ into the string of 18 codes

1101|100|00|101|01|11100|00|00|01|101|1101|101|100|101|00|01|11100|100

The result is a 52-bit string that compresses the original 59 bits. There is almost no compression because $p$ isn't large. Notice that string (2.11) has three short runs of 1's, which can be interpreted as four empty runs of zeros. It also has three runs (of zeros) of length 1. The next example is the 94-bit string

00000000001000000000100000001000000000001000000001000000000001000000001000000100000000010000000,

which is sparser and therefore compresses better. It consists of 85 zeros and 9 ones, so $p = 85/(85 + 9) = 0.9$. The best value of $m$ is therefore $m = \lceil -\log(1.9)/\log(0.9)\rceil = \lceil 6.09\rceil = 7$. The 10 runs of zeros have lengths 10, 9, 7, 11, 8, 12, 8, 7, 10, and 7. When encoded by the Golomb codes for $m = 7$, the run lengths become the 47-bit string

$$10100|10011|1000|10101|10010|10110|10010|1000|10100|1000,$$

resulting in a compression factor of $94/47 = 2$.

The third, extreme, example is a really sparse binary string that consists of, say, $10^6$ bits, of which only 100 are ones. The probability of zero is $p = 10^6/(10^6 + 10^2) = 0.9999$, implying $m = 6932$. There are 101 runs, each about $10^4$ zeros long. The Golomb code of $10^4$ for $m = 6932$ is 14 bits long, so the 101 runs can be compressed to 1414 bits, yielding the impressive compression factor of 707!

In summary, given a binary string, we can employ the method of run-length encoding to compress it with Golomb codes in the following steps: (1) count the number of zeros and ones, (2) compute the probability $p$ of a zero, (3) use Equation (2.14) to compute $m$, (4) construct the family of Golomb codes for $m$, and (5) for each run-length of $n$ zeros, write the Golomb code of $n$ on the compressed stream.

In order for the run lengths to be meaningful, $p$ should be large. Small values of $p$, such as 0.1, result in a string with more 1's than zeros and thus in many short runs of zeros and long runs of 1's. In such a case, it is possible to use RLE to compress the runs of 1's. In general, we can talk about a binary string whose elements are $r$ and $s$ (for run and stop). For $r$, we should select the more common element, but it has to be very common (the distribution of $r$ and $s$ should be skewed) for RLE to produce good compression. Values of $p$ around 0.5 result in runs of both zeros and 1's, so regardless of which bit is selected for the $r$ element, there will be many runs of length zero. For example, the string 000111001100001111101000111 has the following run lengths of zeros 3, 0, 0, 2, 0, 4, 0, 0, 0, 1, 3, 0, 0 and similar run lengths of 1's 0, 0, 3, 0, 2, 0, 0, 0, 4, 1, 0, 0, 3. In such a case, RLE is not a good choice for compression and other methods should be considered.

Another approach to adaptive RLE is to use the binary string input so far to estimate $p$ and from it to estimate $m$, and then use the new value of $m$ to encode the next run length (not the current one because the decoder cannot mimic this). Imagine that three runs of 10, 15, and 21 zeros have been input so far, and the first two have already been compressed. The current run of 21 zeros is first compressed with the current value of $m$, then a new $p$ is computed as $(10 + 15 + 21)/[(10 + 15 + 21) + 3]$

and is used to update $m$ either from $-1/\log_2 p$ or from Equation (2.14). (The 3 is the number of 1's input so far, one for each run.) The new $m$ is used to compress the next run. The algorithm accumulates the lengths of the runs in variable $L$ and the number of runs in $N$. Figure 2.49 is a simple pseudocode listing of this method. (A practical implementation should halve the values of $L$ and $N$ from time to time, to prevent them from overflowing.)

```
L = 0; % initialize
N = 0;
m = 1; % or ask user for m
% main loop
for each run of r zeros do
    construct Golomb code for r using current m.
    write it on compressed stream.
    L = L + r:              % update L, N, and m
    N = N + 1;
    p = L/(L + N);
    m = ⌊-1/ log₂ p + 0.5⌋;
endfor;
```

Figure 2.49: Simple Adaptive Golomb RLE Encoding.

In addition to the codes, Solomon W. Golomb has his "own" Golomb constant: 0.62432998854355087099293638310083724417964262018052928 6.

# 2.24 Rice Codes

The Golomb codes constitute a family that depends on the choice of a parameter $m$. The case where $m$ is a power of 2 ($m = 2^k$) is special and results in a Rice code (sometimes also referred to as Golomb–Rice code), so named after its originator, Robert F. Rice ([Rice 79], [Rice 91], and [Fenwick 96]). The Rice codes are also related to the subexponential code of Section 2.25. A Rice code depends on the choice of a base $k$ and is computed in the following steps: (1) Separate the sign bit from the rest of the number. This is optional and the bit becomes the most-significant bit of the Rice code. (2) Separate the $k$ LSBs. They become the LSBs of the Rice code. (3) Code the remaining $j = \lfloor n/2^k \rfloor$ bits as either $j$ zeros followed by a 1 or $j$ 1's followed by a 0 (similar to the unary code). This becomes the middle part of the Rice code. Thus, this code is computed with a few logical operations, which is faster than computing a Huffman code, a process that requires sliding down the Huffman tree while collecting the individual bits of the code. This feature is especially important for the decoder, which has to be simple and fast. Table 2.50 shows examples of this code for $k = 2$, which corresponds to $m = 4$ (the column labeled "No. of ones" lists the number of 1's in the middle part of the code). Notice that the codes of this table are identical (except

| $i$ | Binary | Sign | LSB | No. of ones | Code | $i$ | Code |
|---|---|---|---|---|---|---|---|
| 0 | 0 | 0 | 00 | 0 | 0\|0\|00 | | |
| 1 | 1 | 0 | 01 | 0 | 0\|0\|01 | -1 | 1\|0\|01 |
| 2 | 10 | 0 | 10 | 0 | 0\|0\|10 | -2 | 1\|0\|10 |
| 3 | 11 | 0 | 11 | 0 | 0\|0\|11 | -3 | 1\|0\|11 |
| 4 | 100 | 0 | 00 | 1 | 0\|10\|00 | -4 | 1\|10\|00 |
| 5 | 101 | 0 | 01 | 1 | 0\|10\|01 | -5 | 1\|10\|01 |
| 6 | 110 | 0 | 10 | 1 | 0\|10\|10 | -6 | 1\|10\|10 |
| 7 | 111 | 0 | 11 | 1 | 0\|10\|11 | -7 | 1\|10\|11 |
| 8 | 1000 | 0 | 00 | 2 | 0\|110\|00 | -8 | 1\|110\|00 |
| 11 | 1011 | 0 | 11 | 2 | 0\|110\|11 | -11 | 1\|110\|11 |
| 12 | 1100 | 0 | 00 | 3 | 0\|1110\|00 | -12 | 1\|1110\|00 |
| 15 | 1111 | 0 | 11 | 3 | 0\|1110\|11 | -15 | 1\|1110\|11 |

Table 2.50: Various Positive and Negative Rice Codes.

for the extra, optional, sign bit) to the codes on the 3rd row (the row for $m = 4$) of Table 2.46.

The length of the (unsigned) Rice code of the integer $n$ with parameter $k$ is $1 + k + \lfloor n/2^k \rfloor$ bits, indicating that these codes are suitable for data where the integer $n$ appears with a probability $P(n)$ that satisfies $\log_2 P(n) = -(1 + k + n/2^k)$ or $P(n) \propto 2^{-n}$, an exponential distribution, such as the Laplace distribution. (See [Kiely 04] for a detailed analysis of the optimal values of the Rice parameter.) The Rice code is instantaneous, once the decoder reads the sign bit and skips to the first 0 from the left, it knows how to generate the left and middle parts of the code. The next $k$ bits should be read and appended to that.

Rice codes are ideal for data items with a Laplace distribution, but other prefix codes exist that are easier to construct and to decode and that may, in certain circumstances, outperform the Rice codes. Table 2.51 lists three such codes. The "pod" code, due to Robin Whittle [firstpr 06], codes the number 0 with the single bit 1, and codes the binary number $1\underbrace{b...b}_{k}$ as $\underbrace{0...0}_{k+1}1\underbrace{b...b}_{k}$. In two cases, the pod code is one bit longer than the Rice code, in four cases it has the same length, and in all other cases it is shorter than the Rice codes. The Elias gamma code [Fenwick 96] is identical to the pod code minus its leftmost zero. It is therefore shorter, but does not provide a code for zero (see also Table 2.8). The biased Elias gamma code corrects this fault in an obvious way, but at the cost of making some codes one bit longer.

There remains the question of what base value $n$ to select for the Rice codes. The base determines how many low-order bits of a data symbol are included directly in the Rice code, and this is linearly related to the variance of the data symbol. Tony Robinson, the developer of the Shorten method for audio compression [Robinson 94], provides the formula $n = \log_2[\log(2)E(|x|)]$, where $E(|x|)$ is the expected value of the data symbols. This value is the sum $\sum |x|p(x)$ taken over all possible symbols $x$.

Figure 2.52 lists the lengths of various Rice codes and compares them to the length of the standard binary (beta) code.

| Number | | Pod | Elias | Biased Elias |
| Dec | Binary | | gamma | gamma |
|---|---|---|---|---|
| 0 | 00000 | 1 | | 1 |
| 1 | 00001 | 01 | 1 | 010 |
| 2 | 00010 | 0010 | 010 | 011 |
| 3 | 00011 | 0011 | 011 | 00100 |
| 4 | 00100 | 000100 | 00100 | 00101 |
| 5 | 00101 | 000101 | 00101 | 00110 |
| 6 | 00110 | 000110 | 00110 | 00111 |
| 7 | 00111 | 000111 | 00111 | 0001000 |
| 8 | 01000 | 00001000 | 0001000 | 0001001 |
| 9 | 01001 | 00001001 | 0001001 | 0001010 |
| 10 | 01010 | 00001010 | 0001010 | 0001011 |
| 11 | 01011 | 00001011 | 0001011 | 0001100 |
| 12 | 01100 | 00001100 | 0001100 | 0001101 |
| 13 | 01101 | 00001101 | 0001101 | 0001110 |
| 14 | 01110 | 00001110 | 0001110 | 0001111 |
| 15 | 01111 | 00001111 | 0001111 | 000010000 |
| 16 | 10000 | 0000010000 | 000010000 | 000010001 |
| 17 | 10001 | 0000010001 | 000010001 | 000010010 |
| 18 | 10010 | 0000010010 | 000010010 | 000010011 |

Table 2.51: Pod, Elias Gamma, and Biased Elias Gamma Codes.

# 2.25 Subexponential Code

The subexponential code of this section is related to the Rice codes. Like the Golomb codes and the Rice codes, the subexponential code depends on a parameter $k \geq 0$. The main feature of the subexponential code is its length. For integers $n < 2^{k+1}$, the code length increases linearly with $n$, but for larger values of $n$, it increases logarithmically. The subexponential code of the nonnegative integer $n$ is computed in two steps. In the first step, values $b$ and $u$ are calculated by

$$b = \begin{cases} k, & \text{if } n < 2^k, \\ \lfloor \log_2 n \rfloor, & \text{if } n \geq 2^k, \end{cases} \quad \text{and} \quad u = \begin{cases} 0, & \text{if } n < 2^k, \\ b - k + 1, & \text{if } n \geq 2^k. \end{cases}$$

In the second step, the unary code of $u$ (in $u+1$ bits) is followed by the $b$ least-significant bits of $n$ to become the subexponential code of $n$. Thus, the total length of the code is

$$u + 1 + b = \begin{cases} k + 1, & \text{if } n < 2^k, \\ 2\lfloor \log_2 n \rfloor - k + 2, & \text{if } n \geq 2^k. \end{cases}$$

Table 2.53 lists examples of the subexponential code for various values of $n$ and $k$. It can be shown that for a given $n$, the code lengths for consecutive values of $k$ differ by at most 1.

Subexponential codes are used in the progressive FELICS method for image compression [Howard and Vitter 94].

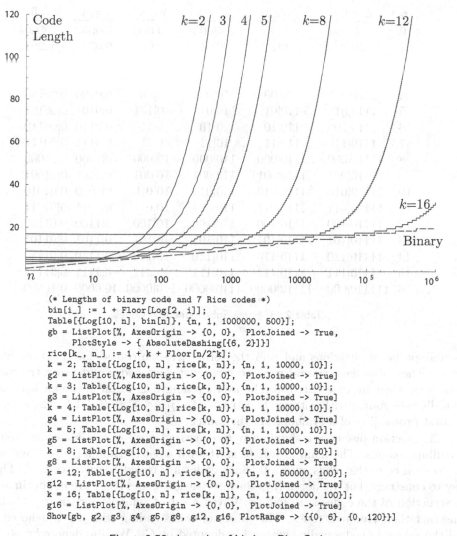

Figure 2.52: Lengths of Various Rice Codes.

The code listing within the figure reads:

```
(* Lengths of binary code and 7 Rice codes *)
bin[i_] := 1 + Floor[Log[2, i]];
Table[{Log[10, n], bin[n]}, {n, 1, 1000000, 500}];
gb = ListPlot[%, AxesOrigin -> {0, 0}, PlotJoined -> True,
        PlotStyle -> { AbsoluteDashing[{6, 2}]}]
rice[k_, n_] := 1 + k + Floor[n/2^k];
k = 2; Table[{Log[10, n], rice[k, n]}, {n, 1, 10000, 10}];
g2 = ListPlot[%, AxesOrigin -> {0, 0}, PlotJoined -> True]
k = 3; Table[{Log[10, n], rice[k, n]}, {n, 1, 10000, 10}];
g3 = ListPlot[%, AxesOrigin -> {0, 0}, PlotJoined -> True]
k = 4; Table[{Log[10, n], rice[k, n]}, {n, 1, 10000, 10}];
g4 = ListPlot[%, AxesOrigin -> {0, 0}, PlotJoined -> True]
k = 5; Table[{Log[10, n], rice[k, n]}, {n, 1, 10000, 10}];
g5 = ListPlot[%, AxesOrigin -> {0, 0}, PlotJoined -> True]
k = 8; Table[{Log[10, n], rice[k, n]}, {n, 1, 100000, 50}];
g8 = ListPlot[%, AxesOrigin -> {0, 0}, PlotJoined -> True]
k = 12; Table[{Log[10, n], rice[k, n]}, {n, 1, 500000, 100}];
g12 = ListPlot[%, AxesOrigin -> {0, 0}, PlotJoined -> True]
k = 16; Table[{Log[10, n], rice[k, n]}, {n, 1, 1000000, 100}];
g16 = ListPlot[%, AxesOrigin -> {0, 0}, PlotJoined -> True]
Show[gb, g2, g3, g4, g5, g8, g12, g16, PlotRange -> {{0, 6}, {0, 120}}]
```

## 2.26 Codes Ending with "1"

In general, the particular bits that constitute a code are irrelevant. Given a set of codewords that have the desired properties, we don't check the individual bits of a code and object if there is a dearth of zeros or too few 1's. Similarly, we don't complain if most or all of the codes start with a 1 or end with a 0. The important requirements of variable-length codes are (1) to have a set of codes that feature the shortest average length for a given statistical distribution of the source symbols and (2) to have a UD code. However, there may be applications where it is advantageous to have codes that start or end in a special way, and this section presents prefix codes that end with a 1. The main contributors to this line of research are R. Capocelli, A. De Santis, T. Berger, and R. Yeung (see [Capocelli and De Santis 94] and [Berger and Yeung 90]). They

| $n$ | $k=0$ | $k=1$ | $k=2$ | $k=3$ | $k=4$ | $k=5$ |
|---|---|---|---|---|---|---|
| 0 | 0| | 0|0 | 0|00 | 0|000 | 0|0000 | 0|00000 |
| 1 | 10| | 0|1 | 0|01 | 0|001 | 0|0001 | 0|00001 |
| 2 | 110|0 | 10|0 | 0|10 | 0|010 | 0|0010 | 0|00010 |
| 3 | 110|1 | 10|1 | 0|11 | 0|011 | 0|0011 | 0|00011 |
| 4 | 1110|00 | 110|00 | 10|00 | 0|100 | 0|0100 | 0|00100 |
| 5 | 1110|01 | 110|01 | 10|01 | 0|101 | 0|0101 | 0|00101 |
| 6 | 1110|10 | 110|10 | 10|10 | 0|110 | 0|0110 | 0|00110 |
| 7 | 1110|11 | 110|11 | 10|11 | 0|111 | 0|0111 | 0|00111 |
| 8 | 11110|000 | 1110|000 | 110|000 | 10|000 | 0|1000 | 0|01000 |
| 9 | 11110|001 | 1110|001 | 110|001 | 10|001 | 0|1001 | 0|01001 |
| 10 | 11110|010 | 1110|010 | 110|010 | 10|010 | 0|1010 | 0|01010 |
| 11 | 11110|011 | 1110|011 | 110|011 | 10|011 | 0|1011 | 0|01011 |
| 12 | 11110|100 | 1110|100 | 110|100 | 10|100 | 0|1100 | 0|01100 |
| 13 | 11110|101 | 1110|101 | 110|101 | 10|101 | 0|1101 | 0|01101 |
| 14 | 11110|110 | 1110|110 | 110|110 | 10|110 | 0|1110 | 0|01110 |
| 15 | 11110|111 | 1110|111 | 110|111 | 10|111 | 0|1111 | 0|01111 |
| 16 | 111110|0000 | 11110|0000 | 1110|0000 | 110|0000 | 10|0000 | 0|10000 |

Table 2.53: Some Subexponential Codes.

discuss special applications and coin the term "feasible codes" for this type of prefix codes. They also show several ways to construct such codes and prove the following bounds on their average lengths. The average length $E$ of an optimum feasible code for a discrete source with entropy $H$ satisfies $H + p_N \leq E \leq H + 1.5$, where $p_N$ is the smallest probability of a symbol from the source.

This section discusses a simple way to construct a set of feasible codes from a set of Huffman codes. The main idea is to start with a set of Huffman codes and append a 1 to each code that ends with a 0. Such codes are called "derived codes." They are easy to construct, but are not always the best feasible codes (they are not optimal). The construction of the codes is as follows: Given a set of symbols and their probabilities, construct their Huffman codes. The subset of codes ending with a 0 is denoted by $C_0$ and the subset of codes ending with a 1 is denoted by $C_1$. We also denote by $p(C_0)$ and $p(C_1)$ the sum of probabilities of the symbols in the two subsets, respectively. Notice that $p(C_0) + p(C_1) = 1$, which implies that either $p(C_0)$ or $p(C_1)$ is less than or equal to $1/2$ and the other one is greater than or equal to $1/2$. If $p(C_0) \leq 1/2$, the derived code is constructed by appending a 1 to each codeword in $C_0$; the codewords in $C_1$ are not modified. If, on the other hand, $p(C_0) > 1/2$, then the zeros and 1's are interchanged in the entire set of Huffman codes, resulting in a new $p(C_0)$ that is less than or equal to $1/2$, and the derived code is constructed as before.

As an example, consider the set of six symbols with probabilities 0.26, 0.24, 0.14, 0.13, 0.12, and 0.11. The entropy of this set is easily computed at approximately 2.497 and one set of Huffman codes for these symbols (from high to low probabilities) is 11, 01, 101, 100, 001, and 000. The average length of this set is

$$0.26 \times 2 + 0.24 \times 2 + 0.14 \times 3 + 0.13 \times 3 + 0.12 \times 3 + 0.11 \times 3 = 2.5 \text{ bits.}$$

It is easy to verify that $p(C_0) = 0.13 + 0.11 = 0.24 < 0.5$, so the derived code becomes 11, 01, 101, 1001, 001, and 0001, with an average size of 2.74 bits. On the other hand, if we consider the set of Huffman codes 00, 10, 010, 011, 110, and 111 for the same symbols, we compute $p(C_0) = 0.26 + 0.24 + 0.14 + 0.12 = 0.76 > 0.5$, so we have to interchange the bits and append a 1 to the codes of 0.13 and 0.11. This again results in a derived code with an average length $E$ of 2.74, which satisfies $E < H + 1.5$.

It is easy to see how the extra 1's added to the codes increase its average size. Suppose that subset $C_0$ includes codes 1 through $k$. Originally, the average length of these codes was $E_0 = p_1 l_1 + p_2 l_2 + \cdots + p_k l_k$ where $l_i$ is the length of code $i$. After a 1 is appended to each of the $k$ codes, the new average length is

$$E = p_1(l_1 + 1) + p_2(l_2 + 1) + \cdots + p_k(l_k + 1) = E_0 + (p_1 + p_2 + \cdots + p_k) \le E_0 + 1/2.$$

The average length has increased by less than half a bit.

Any sufficiently advanced bug is indistinguishable from a feature.
—Arthur C. Clarke, paraphrased by Rich Kulawiec

# 3
# Robust Codes

The many codes included in this chapter have a common feature; thet are robust. Any errors that creep into a string of such codes either can be detected (or even corrected automatically) or have only limited effects and do not propagate indefinitely. The chapter starts with a discussion of the principles of error-control codes.

## 3.1 Codes For Error Control

When data is stored or transmitted, it is often encoded. Modern data communications is concerned with the following types of codes:

- Reliability. The detection and removal of errors caused by noise in the communications channel (this is also referred to as channel coding or error-control coding).

- Efficiency. The efficient encoding of the information in a small number of bits (source coding or data compression).

- Security. Protecting data against eavesdropping, intrusion, or tampering (the fields of cryptography and data hiding).

This section and the next one cover the chief aspects of error-control as applied to fixed-length codes. Section 3.3 returns to variable-length codes and the remainder of this chapter is concerned with the problem of constructing reliable (or robust) codes.

Every time information is transmitted, over any channel, it may get corrupted by noise. In fact, even when information is stored in a storage device, it may become bad, because no piece of hardware is absolutely reliable. This important fact also applies to non-computer information. Text written or printed on paper fades over time and the paper itself degrades and may crumble. Audio and video data recorded on magnetic media fades over time. Speech sent on the air becomes corrupted by noise, wind, and fluctuations of temperature and humidity. Speech, in fact, is a good starting point for understanding the principles of error-control. Imagine a noisy cocktail party where everybody talks simultaneously, on top of blaring music. We know that even in such a situation it is possible to carry on a conversation, but more attention than normal is needed.

What makes our language so robust, so immune to errors, are two properties, redundancy and context.

■  Our language is redundant because only a very small fraction of all possible words are valid. A huge number of words can be constructed with the 26 letters of the English alphabet. Just the number of 7-letter words, for example, is $26^7 \approx 8.031$ billion. Yet only about 50,000 words are commonly used in daily conversations, and even the Oxford English Dictionary lists "only" about 500,000 words. When we hear a garbled word, our brain searches through many similar words for the "closest" valid word. Computers are very good at such searches, which is why redundancy is the basis of error-control codes.

■  Our brain works by associations. This is why we humans excel at using the context of a message to locate and correct errors in the message. In receiving a sentence with a garbled word or a word that doesn't belong, such as "pass the thustard please," we first search our memory to find words that are associated with "thustard," then we use our accumulated life experience to select, among perhaps many possible candidates, the word that best fits in the present context. If we are driving on the highway, we pass the bastard in front of us; if we are at dinner, we pass the mustard (or custard). Another example is the (corrupted) written sentence a*l n*tu*al l**gua*es a*e red***ant, which we can easily complete. Computers don't have much life experience and are notoriously bad at such tasks, which is why context is not used in digital codes. In extreme cases, where much of the sentence is bad, even we may not be able to correct it, and we have to ask for a retransmission "say it again, Sam."

The idea of using redundancy to add reliability to information is due to Claude Shannon, the founder of information theory. It is not an obvious idea, since we are conditioned against it. Most of the time, we try to *eliminate* redundancy in digital data, in order to save space and compress the data. This habit is also reflected in our everyday, nondigital behavior. We shorten Michael to Mike and Samuel to Sam, and it is a rare Bob who insists on being a Robert.

There are several approaches to robust codes, but this section deals only with the concept of *check bits*, because this leads to the important concept of *Hamming distance*. Section 3.7 shows how this concept can be extended to variable-length codes.

Imagine a text message encoded in $m$-bit words (perhaps ASCII or Unicode) and then stored or transmitted. We can make the message robust by adding several check bits to the $m$ data bits of each word of the message. We assume that $k$ check bits are appended to the original $m$ information bits, to produce a codeword of $n = m + k$ bits. Such a code is referred to as an $(n, m)$ code. The codewords are then transmitted and decoded at the receiving end. Only certain combinations of the information bits and check bits are valid, in analogy with a natural language. The decoder knows what the valid codewords are. If a nonvalid codeword is received, the decoder considers it an error. By adding more check bits, the decoder can also correct certain errors, not just detect them. The principle of error correction, not just detection, is that, on receiving a bad codeword, the receiver selects the valid codeword that is the "closest" to it.

**Example**: Assume a set of 128 symbols (i.e., $m = 7$). If we select $k = 4$, we end up with 128 valid codewords, each 11 bits long. This is an $(11, 7)$ code. The valid codewords are selected from a total of $2^{11} = 2048$ possible codewords, so there remain $2048 - 128 = 1920$ nonvalid codewords. The big difference between the number of valid (128) and nonvalid (1920) codewords implies that if a codeword gets corrupted, chances are that it will change to a nonvalid one.

It may, of course, happen that a valid codeword gets modified, during transmission, to another valid codeword. Thus, our codes are not absolutely reliable, but can be made more and more robust by adding more check bits and by selecting the valid codewords carefully. The noisy channel theorem, one of the basic theorems of information theory, states that codes can be made as reliable as desired, even in the presence of much noise, by adding check bits and thus separating our codewords further and further, as long as the rate of transmission does not exceed a certain quantity referred to as the channel's capacity.

It is important to understand the meaning of the word "error" in data processing and communications. When an $n$-bit codeword is transmitted over a channel, the decoder may receive the same $n$ bits, it may receive $n$ bits, some of which are bad, but it may also receive fewer than or more than $n$ bits. Thus, bits may be added, deleted, or changed (substituted) by noise in the communications channel. In this section we consider only substitution errors. A bad bit simply changes its value, either from 0 to 1, or from 1 to 0. This makes it relatively easy to correct the bit. If the error-control code tells the receiver which bits are bad, the receiver corrects those bits by inverting them.

Parity bits represent the next step in error-control. A parity bit can be added to a group of $m$ information bits to complete the total number of 1's to an odd number. Thus, the (odd) parity of the group 10110 is 0, since the original group plus the parity bit has an odd number (3) of 1's. It is also possible to use even parity, and the only difference between odd and even parity is that, in the case of even parity, a group of all zeros is valid, whereas, with odd parity, any group of bits with a parity bit added, cannot be all zeros.

Parity bits can be used to design simple, but not very efficient, error-correcting codes. To correct 1-bit errors, the message can be organized as a rectangle of dimensions $(r - 1) \times (s - 1)$. A parity bit is added to each row of $s - 1$ bits, and to each column of $r - 1$ bits. The total size of the message (Table 3.1) is now $s \times r$.

```
0 1 0 0 1
1 0 1 0 0                    0 1 0 0 1
0 1 1 1 1                    1 0 1 0
0 0 0 0 0                    0 1 0
1 1 0 1 1                    0 0
0 1 0 0 1                    1
```

Table 3.1.                    Table 3.2.

If only one bit becomes bad, a check of all $s - 1 + r - 1$ parity bits will detect the error, since only one of the $s - 1$ parities and only one of the $r - 1$ parities will be bad.

The overhead of a code is defined as the number of parity bits divided by the number of information bits. The overhead of the rectangular code is, therefore,

$$\frac{(s - 1 + r - 1)}{(s - 1)(r - 1)} \approx \frac{s + r}{s \times r - (s + r)}.$$

A similar, slightly more efficient code is a triangular configuration, where the information bits are arranged in a triangle, with the parity bits placed on the diagonal (Table 3.2). Each parity bit is the parity of all the bits in its row and column. If the top row contains $r$ information bits, the entire triangle has $r(r + 1)/2$ information bits and $r$ parity bits. The overhead is therefore

$$\frac{r}{r(r + 1)/2} = \frac{2}{r + 1}.$$

It is also possible to arrange the information bits in a number of two-dimensional planes, to obtain a three-dimensional cube, three of whose six outer surfaces consist of parity bits.

It is not obvious how to generalize these methods to more than 1-bit error correction.

**Hamming Distance and Error Detecting.** In the 1950s, Richard Hamming conceived the concept of distance as a general way to use check bits for error detection and correction.

| Symbol | code$_1$ | code$_2$ | code$_3$ | code$_4$ | code$_5$ |
|--------|------|------|------|--------|--------|
| $A$ | 0000 | 0000 | 001 | 001001 | 01011 |
| $B$ | 1111 | 1111 | 010 | 010010 | 10010 |
| $C$ | 0110 | 0110 | 100 | 100100 | 01100 |
| $D$ | 0111 | 1001 | 111 | 111111 | 10101 |
| $k$: | 2 | 2 | 1 | 4 | 3 |

Table 3.3: Codes with $m = 2$.

To illustrate this idea, we start with a simple example of four symbols $A$, $B$, $C$, and $D$. Only two information bits are required, but the codes of Table 3.3 add some check bits, for a total of 3–6 bits per symbol. The first of these codes, code$_1$, is simple. Its four codewords were selected from the 16 possible 4-bit numbers, and are not the best possible ones. When the receiver receives one of them, say, 0111, it assumes that there is no error and the symbol received is $D$. When a nonvalid codeword is received, the receiver signals an error. Since code$_1$ is not the best possible, not every error is detected. Even if we limit ourselves to single-bit errors, this code is not very good. There are 16 possible single-bit errors in its four 4-bit codewords, and of those, the following four cannot be detected: a 0110 changed during transmission to 0111, a 0111 modified to 0110, a 1111 corrupted to 0111, and a 0111 changed to 1111. Thus, the error detection rate is 12 out of 16, or 75%. In contrast, code$_2$ does a much better job. It can detect every single-bit error, because when only a single bit is changed in any of its codewords, the result is not any of the other codewords. We say that the four codewords of code$_2$ are sufficiently distant from one another. The concept of distance of codewords is easy to describe.

1. Two codewords are a Hamming distance $d$ apart if they differ in exactly $d$ of their $n$ bits.

2. A code has a Hamming distance of $d$ if every pair of codewords in the code is at least a Hamming distance $d$ apart.

(For mathematically-inclined readers.) These definitions have a simple geometric interpretation. Imagine a hypercube in $n$-dimensional space. Each of its $2^n$ corners can be numbered by an $n$-bit number (Figure 3.4) such that each of the $n$ bits corresponds to one of the $n$ dimensions of the cube. In such a cube, points that are directly connected (near neighbors) have a Hamming distance of 1, points with a common neighbor have a Hamming distance of 2, and so on. If a code with a Hamming distance of 2 is desired, only points that are not directly connected should be selected as valid codewords.

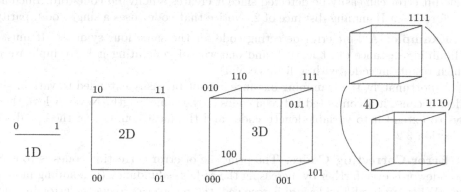

Figure 3.4: Cubes of Various Dimensions and Corner Numbering.

The "distance" approach to reliable communications is natural and has been employed for decades by many organizations—most notably armies, but also airlines and emergency services—in the form of phonetic alphabets. We know from experience that certain letters—such as F and S, or B, D, and V—sound similar over the telephone and often cause misunderstanding in verbal communications (the pair 8 and H also comes to mind). A phonetic alphabet replaces each letter by a word that starts with that letter, selecting words that are "distant" in some sense, in order to remove any ambiguity.

Thus, replacing F by foxtrot and S by sierra guarantees reliable conversations, because these words sound very different and will not be mistaken even under conditions of noise and garbled communications.

The NATO phonetic alphabet is Alfa Bravo Charlie Delta Echo Foxtrot Golf Hotel India Juliett Kilo Lima Mike November Oscar Papa Quebec Romeo Sierra Tango Uniform Victor Whiskey X-ray Yankee Zulu. For other phonetic alphabets, see [uklinux 07]. Reference [codethatword 07] may also be of interest to some.

The reason code$_2$ can detect all single-bit errors is that it has a Hamming distance of 2. The distance between valid codewords is 2, so a 1-bit error always changes a valid codeword into a nonvalid one. When two bits go bad, a valid codeword is moved to another codeword at distance 2. If we want that other codeword to be nonvalid, the code must have at least distance 3.

In general, a code with a Hamming distance of $d + 1$ can detect all $d$-bit errors. In comparison, code$_3$ has a Hamming distance of 2 and can therefore detect all 1-bit errors even though it is short ($n = 3$). Similarly, code$_4$ has a Hamming distance of 4, which is more than enough to detect all 2-bit errors. It is now obvious that we can increase the reliability of our data, but this feature does not come free. As always, there is a tradeoff, or a price to pay, in the form of overhead. Our codes are much longer than $m$ bits per symbol because of the added check bits. A measure of the price is $n/m = \frac{m+k}{m} = 1 + k/m$, where the quantity $k/m$ is the overhead of the code. In the case of code$_1$ the overhead is 2, and in the case of code$_3$ it is 3/2.

**Example:** A code with a single check bit, that is a parity bit (even or odd). Any single-bit error can easily be detected since it creates a nonvalid codeword. Such a code therefore has a Hamming distance of 2. Notice that code$_3$ uses a single, odd, parity bit.

**Example:** A 2-bit error-detecting code for the same four symbols. It must have a Hamming distance of at least 3, and one way of generating it is to duplicate code$_3$ (which results in code$_4$ with a distance of 4).

Unfortunately, the Hamming distance cannot be easily extended to variable-length codes, because it is computed between codes of the same length. Nevertheless, there are ways to extend it to variable-length codes and the most common of these is discussed in Section 3.2.

**Error-Correcting Codes.** The principle of error-correcting codes is to separate the codewords even farther by increasing the code's redundancy (i.e., adding more check bits). When an invalid codeword is received, the receiver corrects the error by selecting the valid codeword that is closest to the one received. An example is code$_5$, which has a Hamming distance of 3. When one bit is modified in any of its four codewords, that codeword is one bit distant from the original, but is still two bits distant from any of the other codewords. Thus, if there is only one error, the receiver can always correct it.

In general, when $d$ bits go bad in a codeword $C_1$, it turns into an invalid codeword $C_2$ at a distance $d$ from $C_1$. If the distance between $C_2$ and the other valid codewords is at least $d + 1$, then $C_2$ is closer to $C_1$ than it is to any other valid codeword. This is why a code with a Hamming distance of $d + (d + 1) = 2d + 1$ can correct all $d$-bit errors.

How are the codewords selected? The problem is to select a good set of $2^m$ codewords out of the $2^n$ possible ones. The simplest approach is to use brute force. It is easy to write a computer program that will examine all the possible sets of $2^m$ codewords, and will stop at the first one that has the right distance. The problems with this approach are (1) the time and storage required at the receiving end to verify and correct the codes received, and (2) the amount of time it takes to examine all the possibilities.

**Problem 1.** The receiver must have a list of all the $2^n$ possible codewords. For each codeword, it must have a flag indicating whether it is valid, and if not, which valid codeword is the closest to it. Every codeword received has to be searched for and found in this list in order to verify it.

**Problem 2**. In the case of four symbols, only four codewords need be selected. For code$_1$ and code$_2$, these four codewords had to be selected from among 16 possible numbers, which can be done in $\binom{16}{4} = 7280$ ways. It is possible to write a simple program that will systematically select sets of four codewords until it finds a set with the required distance. In the case of code$_4$, however, the four codewords had to be selected from a set of 64 numbers, which can be done in $\binom{64}{4} = 635{,}376$ ways. It is still possible to write a program that will systematically explore all the possible codeword selections. In practical cases, however, where sets of hundreds of symbols are involved, the number of possibilities of selecting codewords is too large even for the fastest computers to handle in reasonable time.

This is why sophisticated methods are needed to construct sets of error-control codes. Such methods are out of the scope of this book but are discussed in many books and publications on error control codes. Section 3.7 discusses approaches to developing robust variable-length codes.

# 3.2 The Free Distance

The discussion above shows that the Hamming distance is an important metric (or measure) of the reliability (robustness) of an error-control code. This section shows how the concept of distance can be extended to variable-length codes. Given a code with $s$ codewords $c_i$, we first construct the length vector of the code. We assume that there are $s_1$ codewords of length $L_1$, $s_2$ codewords of length $L_2$, and so on, up to $s_m$ codewords of length $L_m$. We also assume that the lengths $L_i$ are sorted such that $L_1$ is the shortest length and $L_m$ is the longest. The length vector is $(L_1, L_2, \ldots, L_m)$.

The first quantity defined is $b_i$, the minimum block distance for length $L_i$. This is simply the minimum Hamming distance of the $s_i$ codewords of length $L_i$ (where $i$ goes from 1 to $m$). The overall minimum block length $b_{\min}$ is defined as the smallest $b_i$.

We next look at all the possible pairs of codewords $(c_i, c_j)$. The two codewords of a pair may have different lengths, so we first compute the distances of their prefixes. Assume that the lengths of $c_i$ and $c_j$ are 12 and 4 bits, respectively, we examine the four leftmost bits of the two, and compute their Hamming distance. The minimum of all these distances, over all possible pairs of codewords, is called the minimum diverging distance of the code and is denoted by $d_{\min}$.

Next, we do the same for the postfixes of the codewords. Given a pair of codewords of lengths 12 and 4 bits, we examine their last (rightmost) four bits and compute their Hamming distance. The smallest of these distances is the minimum converging distance of the code and is denoted by $c_{\min}$.

The last step is more complex. We select a positive integer $N$ and construct all the sequences of codewords whose total length is $N$. If there are many codewords, there may be many sequences of codewords whose total length is $N$. We denote those sequences by $f_1$, $f_2$, and so on. The set of all the $N$-bit sequences $f_i$ is denoted by $F_N$. We compute the Hamming distances of all the pairs $(f_i, f_j)$ in $F_N$ for different $i$ and $j$ and select the minimum distance. We repeat this for all the possible values of $N$ (from 1 to infinity) and select the minimum distance. This last quantity is termed the free distance of the code [Bauer and Hagenauer 01] and is denoted by $d_{\text{free}}$. The free distance of a variable-length code is the single most important parameter determining the robustness of the code. This metric is considered the equivalent of the Hamming distance, and [Buttigieg and Farrell 95] show that it is bounded by

$$d_{\text{free}} \geq \min(b_{\text{min}}, d_{\text{min}} + c_{\text{min}}).$$

# 3.3 Synchronous Prefix Codes

Errors are a fact of life. They are all around us, are found everywhere, and are responsible for many glitches and accidents and for much misery and misunderstanding. Unfortunately, computer data is not an exception to this rule and is not immune to errors. Digital data written on a storage device such as a disk, CD, or DVD is subject to corruption. Similarly, data stored in the computer's internal memory can become bad because of a sudden surge in the electrical voltage, a stray cosmic ray hitting the memory circuits, or an extreme variation of temperature. When binary data is sent over a communications channel, errors may creep up and damage the bits. This is why error-detecting and error-correcting codes (also known as error-control codes or channel codes) are so important and are used in many applications. Data written on CDs and DVDs is made very reliable by including sophisticated codes that detect most errors and can even correct many errors automatically. Data sent over a network between computers is also often protected in this way. However, error-control codes have a serious downside, they work by adding extra bits to the data (parity bits or check bits) and thus increase both the redundancy and the size of the data. In this sense, error-control (data reliability and integrity) is the opposite of data compression. The main goal of compression is to eliminate redundancy from the data, but this inevitably decreases data reliability and opens the door to errors and data corruption.

> We are built to make mistakes, coded for error.
> —Lewis Thomas

The problem of errors in communications (in scientific terms, the problem of noisy communications channels or of sources of noise) is so fundamental, that the first figure in Shannon's celebrated 1948 papers [Shannon 48] shows a source of noise (Figure 3.5).

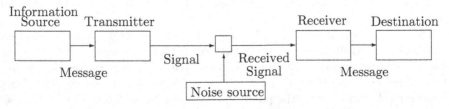

Figure 3.5: The Essence of Communications.

As a result, reliability or lack thereof is the chief downside of variable-length codes. A single error that creeps into a compressed stream can propagate and trigger many consecutive errors when the stream is decompressed. We say that the decoder loses synchronization with the data or that it slips while decoding. The first solution that springs to mind, when we consider this problem, is to add an error-control code to the compressed data. This solves the problem and results in reliable receiving and decoding, but it requires the introduction of many extra bits (often about 50% of the size of the data). Thus, this solution makes sense only in special applications, where data reliability and integrity is more important than size. A more practical solution is to develop *synchronous codes*. Such a code has one or more synchronizing codewords, so it does not artificially increase the size of the compressed data. On the other hand, such a code is not as robust as an error-control code and it allows for a certain amount of slippage for each error. The idea is that an error will propagate and will affect that part of the compressed data from the point it is encountered (from the bad codeword) until the first synchronizing codeword is input and decoded. The corrupted codeword and the few codewords following it (the slippage region) will be decoded into the wrong data symbols, but the synchronizing codeword will synchronize the decoder and stop the propagation of the error. The decoder will recognize either the synchronizing codeword or the codeword that follows it and will produce correct output from that point. Thus, a synchronous code is not as reliable as an error-control code. It allows for a certain amount of error propagation and should be used only in applications where that amount of bad decoded data is tolerable. There is also the important consideration of the average length of a synchronous code. Such a code may be a little longer than other variable-length codes, but it shouldn't be much longer.

The first example shows how a single error can easily propagate through a compressed file that's being decoded and cause a long slippage. We consider the simple feasible code 0001, 001, 0101, 011, 1001, 101, and 11. If we send the string of codewords 000**1**|101|011|101|011|101|011|... and the third bit (in boldface) gets corrupted to a 1, then the decoder will decode this as the string 001|11|0101|11|0101|11|..., resulting in a long slippage.

However, if we send 01**0**1|011|11|101|0001|... and the third bit goes bad, the decoder will produce 011|101|11|11|010001 .... The last string (010001) serves as an error indicator for the decoder. The decoder realizes that there is a problem and the best way for it to resynchronize itself is to skip a bit and try to decode first 10001...(which fails) and then 0001 (a success). The damage is limited to a slippage of a few symbols, because the codeword 0001 is synchronizing. Similarly, 1001 is also a synchronizing codeword, so a string of codewords will synchronize itself after an error when 1001 is read. Thus 01**0**1|011|11|101|1001|... is decoded as 011|101|11|11|011|001..., thereby stopping the slippage.

Slippage: The act or an instance of slipping, especially movement away from an original or secure place.

The next example is the 4-ary code of Table 3.9 (after [Capocelli and De Santis 92]), where the codewords flagged by an "*" are synchronizing. Consider the string of codewords **0**000|100|002|3|... where the first bit is damaged and is input by the decoder as 1. The decoder will generate 100|01|0000|23|..., thereby limiting the slippage to four symbols. The similar string 0000|100|002|23| . . . will be decoded as 100|01|0000|22|3 . . ., and the error in the code fragment 0000|100|002|101| . . . will result in 100|01|0000|21|01 . . . . In all cases, the damage is limited to the substring from the position of the error to the location of the synchronizing codeword.

The material that follows is based on [Ferguson and Rabinowitz 84]. In order to understand how a synchronizing codeword works, we consider a case where an error has thrown the decoder off the right track. If the next few bits in the compressed file are 101110010. . ., then the best thing for the decoder to do is to examine this bit pattern and try to locate a valid codeword in it. The first bit is 1, so the decoder concentrates on all the codewords that start with 1. Of these, the decoder examines all the codewords that start with 10. Of these, it concentrates on all the codewords that starts with 101, and so on. If 101110010. . . does not correspond to any codeword, the decoder discards the first bit and examines the string 01110010. . . in the same way.

Based on this process, it is easy to identify the three main cases that can occur at each decoding step, and through them to figure out the properties that a synchronizing codeword should have. The cases are as follows:

1. The next few bits constitute a valid codeword. The decoder is synchronized and can proceed normally.

2. The next few bits are part of a long codeword $a$ whose suffix is identical to a synchronizing codeword $b$. Suppose that $a = 1|101110010$ and $b = 010$. Suppose also that because of an error the decoder has become unsynchronized and is positioned after the first bit of $a$ (at the vertical bar). When the decoder examines bit patterns as discussed earlier and skips bits, it will eventually arrive at the pattern 010 and identify it as the valid codeword $b$. We know that this pattern is the tail (suffix) of $a$ and is not $b$, but the point is that $b$ has synchronized the decoder and the error no longer propagates. Thus, we know that for $b$ to be a synchronizing codeword, it has to satisfy the following: If $b$ is a substring of a longer codeword $a$, then $b$ should be the suffix of $a$ and should not reappear elsewhere in $a$.

3. The decoder is positioned in front of the string 100|11001 . . . where the 100 is the tail (suffix) of a codeword $x$ that the decoder cannot identify because of a slippage, and 11001 is a synchronizing codeword $c$. Assume that the bit pattern 100|110 (the suffix of $x$ followed by the prefix of $c$) happens to be a valid codeword. If the suffix 01 of $c$ is a valid codeword $a$, then $c$ would do its job and would terminate the slippage (although it wouldn't be identified by the decoder).

Based on these cases, we list the two properties that a codeword should have in order to be synchronizing. (1) If $b$ is a synchronizing codeword and it happens to be the suffix of a longer codeword $a$, then $b$ should not occur anywhere else in $a$. (2) If a prefix of $b$ is identical to a suffix of another codeword $x$, then the remainder of $b$ should be either a valid codeword or identical to several valid codewords. These two properties can be considered the definition of a synchronizing codeword.

Now assume that $C$ is a synchronous code (i.e., a set of codewords, one or more of which are synchronizing) and that $c$ is a synchronizing codeword in $C$. Codeword $c$ is the variable-length code assigned to a symbol $s$ of the alphabet, and we denote the probability of $s$ by $p$. Given a long string of data symbols, $s$ will appear in it $1/p$ percent of the time. Equivalently, every $(1/p)$th symbol will on average be $s$. When such a string is encoded, codeword $c$ will appear with the same frequency and will therefore limit the length of any slippage to a value proportional to $1/p$. If a code $C$ includes $k$ synchronizing codewords, then on average every $(1/\sum_i^k p_i)$th codeword will be a synchronizing codeword. Thus, it is useful to have synchronizing codewords assigned ⁞⁞ ⁞⁞⁞⁞⁞⁞⁞⁞⁞ (⁞⁞⁞⁞ ⁞⁞⁞⁞⁞⁞⁞⁞⁞⁞) ⁞⁞⁞⁞⁞⁞⁞⁞ ⁞⁞⁞ ⁞⁞⁞⁞⁞⁞⁞⁞⁞ ⁞⁞ ⁞⁞⁞⁞⁞⁞⁞⁞⁞⁞⁞ ⁞⁞⁞⁞ ⁞⁞⁞⁞⁞⁞⁞ length codes demands that common symbols be assigned short codewords, leading us to conclude that the best synchronous codes are those that have many synchronizing short codewords. The shortest codewords are single bits, but since 0 and 1 are always suffixes, they cannot satisfy the definition above and cannot serve as synchronizing codewords (however, in a nonbinary code, such as a ternary code, one-symbol codewords can be synchronizing as illustrated in Table 3.9).

Table 3.6 lists four Huffman codes. Code $C_1$ is nonsynchronous and code $C_2$ is synchronous. Figure 3.7 shows the corresponding Huffman code trees. Starting with the data string ADABCDABCDBECAABDECA, we repeat the string indefinitely and encode it in $C_1$. The result starts with 01|000|01|10|11|000|01|10|11|000|10|001|... and we assume that the second bit (in boldface) gets corrupted. The decoder, as usual, inputs the string 00000011011000011011000010001..., proceeds normally, and decodes it as the bit string 000|000|11|01|10|000|11|01|10|001|000|1..., resulting in unbounded slippage. This is a specially-contrived example, but it illustrates the risk posed by a nonsynchronous code.

| Symbol | A | B | C | D | E | | Symbol | A | B | C | D | E | F |
|--------|---|---|---|----|----|--|--------|---|---|----|-----|-----|-----|
| Prob. | 0.3 | 0.2 | 0.2 | 0.2 | 0.1 | | Prob. | 0.3 | 0.3 | 0.1 | 0.1 | 0.1 | 0.1 |
| $C_1$ | 01 | 10 | 11 | 000 | 001 | | $C_3$ | 10 | 11 | 000 | 001 | 010 | 011 |
| $C_2$ | 00 | 10 | 11 | 010 | 011 | | $C_4$ | 01 | 11 | 000 | 001 | 100 | 101 |

Table 3.6: Synchronous and Nonsynchronous Huffman Codes.

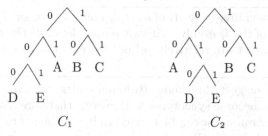

Figure 3.7: Strongly Equivalent Huffman Trees.

In contrast, code $C_2$ is synchronized, with the two synchronizing codewords 010 and 011. Codeword 010 is synchronizing because it satisfies the two parts of the definition above. It satisfies part 1 by default (because it is the longest codeword) and it satisfies part 2 because its suffix 10 is a valid codeword. The argument for 011 is similar. Thus, since both codes are Huffman codes and have the same average code length, code $C_2$ is preferable.

In [Ferguson and Rabinowitz 84], the authors concentrate on the synchronization of Huffman codes. They describe a twisting procedure for turning a nonsynchronous Huffman code tree into an equivalent tree that is synchronous. If the procedure can

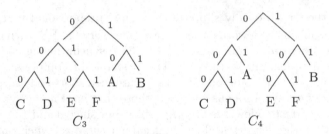

Figure 3.8: Weakly Equivalent Huffman Trees.

be carried out, then the two codes are said to be strongly equivalent. However, this procedure cannot always be performed. Code $C_3$ of Table 3.6 is nonsynchronous, while code $C_4$ is synchronous (with 101 as its synchronizing codeword). The corresponding Huffman code trees are shown in Figure 3.8 but $C_3$ cannot be twisted into $C_4$, which is why these codes are termed weakly equivalent. The same reference also proves the following criteria for nonsynchronous Huffman codes:

1. Given a Huffman code where the codewords have lengths $l_i$, denote by $(l_1, \ldots, l_j)$ the set of all the different lengths. If the greatest common divisor of this set does not equal 1, then the code is nonsynchronous and is not even weakly equivalent to a synchronous code.

2. For any $n \geq 7$, there is a Huffman code with $n$ codewords that is nonsynchronous and is not weakly equivalent to a synchronous code.

3. A Huffman code in which the only codeword lengths are $l$, $l+1$, and $l+3$ for some $l > 2$, is nonsynchronous. Also, for any $n \geq 12$, there is a source where all the Huffman codes are of this type.

> The greatest common divisor (gcd) of a set of nonzero integers $l_i$ is the largest integer that divides each of the $l_i$ evenly (without remainders). If the gcd of a set is 1, then the members of the set are relatively prime. An example of relatively prime integers is $(9, 28, 29)$.

These criteria suggest that many Huffman codes are nonsynchronous and cannot even be modified to become synchronous. However, there are also many Huffman codes that are either synchronous or can be twisted to become synchronous. We start with the concept of quantized source. Given a source $S$ of data symbols with known probabilities, we construct a Huffman code for the source. We denote the length of the longest codeword by $L$ and denote by $\alpha_i$ the number of codewords with length $i$. The vector $S = (\alpha_1, \alpha_2, \ldots, \alpha_L)$ is called the quantized representation of the source $S$. A source is said to be gapless if the first nonzero $\alpha_i$ is followed only by nonzero $\alpha_j$ (i.e., if all the zero $\alpha$'s are concentrated at the start of the quantized vector). The following are criteria for synchronous Huffman codes:

1. If $\alpha_1$ is positive (i.e., there are some codewords with length 1), then any Huffman code for $S$ is synchronous.

2. Given a gapless source $S$ where the minimum codeword length is 2, then $S$ has a synchronous Huffman code, unless its quantized representation is one of the following $(0, 4)$, $(0, 1, 6)$, $(0, 1, 1, 10)$, and $(0, 1, 1, 1, 18)$.

The remainder of this section is based on [Capocelli and De Santis 92], a paper that offers an advanced discussion of synchronous prefix codes, with theorems and proofs. We present only a short summary of the many results discovered by these authors. Given a feasible code (Section 2.26) where the length of the longest codeword is $L$, all the codewords of the form $\underbrace{00\ldots0}_{L-1}1$, $\underbrace{00\ldots0}_{L-2}1$, and $1\underbrace{00\ldots0}_{L-2}1$, are synchronizing codewords. If there are no such codewords, it is possible to modify the code as follows. Select a codeword of length $L$ with the most number of consecutive zeros on the left $\underbrace{00\ldots0}_{j}1x\ldots x$ and change the 1 to a 0 to obtain $\underbrace{00\ldots0}_{j+1}x\ldots x$. Continue in this way, until all the $x$ bits except the last one have been changed to 1's and the codeword has the format $\underbrace{00\ldots0}_{L-1}1$. This codeword is now synchronizing.

As an example, given the feasible code **0001**, 001, 0101, 011, **1001**, 101, and 11, $L = 4$ and the three codewords in boldface are synchronizing. This property can be extended to nonbinary codes, with the difference that instead of a 1, any digit other than 0 can be substituted. Thus, given the feasible ternary code 00001, 0001, 0002, 001, 002, 01, 02, 1001, 101, 102, 201, 21, and 22, $L$ equals 5 and the first three codewords are synchronizing. Going back to the definition of a synchronizing codeword, it is easy to show by direct checks that the first 11 codewords of this code satisfy this definition and are therefore synchronizing.

Table 3.9 (after [Capocelli and De Santis 92]) lists a 4-ary code for the 21-character Latin alphabet. The rules above imply that all the codewords that end with 3 are synchronizing, while our previous definition shows that 101 and 102 are also synchronizing.

| Letter | Freq. | Code | Letter | Freq. | Code | Letter | Freq. | Code |
|--------|-------|------|--------|-------|------|--------|-------|------|
| h | 5 | 0000 | d | 17 | *101 | r | 67 | 20 |
| x | 6 | 0001 | l | 21 | *102 | s | 68 | 21 |
| v | 7 | 0002 | p | 30 | *103 | a | 72 | 22 |
| f | 9 | 100 | c | 33 | 11 | t | 72 | *03 |
| b | 12 | 001 | m | 34 | 12 | u | 74 | *13 |
| q | 13 | 002 | o | 44 | 01 | e | 92 | *23 |
| g | 14 | *003 | n | 60 | 02 | i | 101 | *3 |

Table 3.9: Optimal 4-ary Synchronous Code for the Latin Alphabet.

See also the code of Section 2.20.1.

# 3.4 Resynchronizing Huffman Codes

Because of the importance and popularity of Huffman codes, much research has gone into every aspect of these codes. The work described here, due to [Rudner 71], shows how to identify a resynchronizing Huffman code (RHC), a set of codewords that allows the decoder to always synchronize itself following an error. Such a set contains at least one synchronizing codeword (SC) with the following property: any bit string followed by an SC is a sequence of valid codewords.

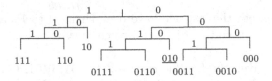

Figure 3.10: A Resynchronizing Huffman Code Tree.

Figure 3.10 is an example of such a code. It is a Huffman code tree for nine symbols having probabilities of occurrence of 1/4, 1/8, 1/8, 1/8, 1/8, 1/16, 1/16, 1/16, and 1/16. Codeword 010 (underlined) is an SC. A direct check verifies that any bit string followed by 010 is a sequence of valid codewords from this tree. For example, 001010001|010 is the string of four codewords 0010|10|0010|10. If an encoded file has an error at point $A$ and an SC appears later, at point $B$, then the bit string from $A$ to $B$ (including the SC) is a string of valid codewords (although not the correct codewords) and the decoder becomes synchronized at $B$ (although it may miss decoding the SC itself). Reference [Rudner 71] proposes an algorithm to construct an RHC, but the lengths of the codewords must satisfy certain conditions.

Given a set of prefix codewords, we denote the length of the shortest code by $m$. An integer $q < m$ is first determined by a complex test (listed below) that goes over all the nonterminal nodes in the code tree and counts the number of codewords of the same length that are descendants of the node. Once $q$ has been computed, the SC is the codeword $0^q 1^{m-q} 0^q$. It is easy to see that $m = 2$ for the code tree of Figure 3.10. The test results in $q = 1$, so the SC is 010.

In order for an SC to exist, the code must either have $m = 1$ (in which case the SC has the form $0^{2q}$) or must satisfy the following conditions:

1. The value of $m$ is 2, 3, or 4.
2. The code must contain codewords with all the lengths from $m$ to $2m - 1$.
3. The value of $q$ that results from the test must be less than $m$.

Figure 3.11 is a bigger example with 20 codewords. The shortest codeword is three bits long, so $m = 3$. The value of $q$ turns out to be 2, so the SC is 00100. A random bit string followed by the SC, such as 01000101100011100101010100010101|00100, becomes the string of nine codewords 0100|01011|00011|100|1010|1010|00101|0100|100.

The test to determine the value of $q$ depends on many quantities that are defined in [Rudner 71] and are too specialized to be described here. Instead, we copy this test verbatim from the reference.

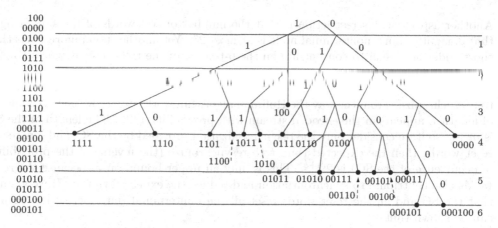

Figure 3.11: A Resynchronizing Huffman Code Tree With 20 Codewords.

"Let index $I$ have $2 \leq m \leq 4$ and $p < m$. Let

$$Q_j(h) = (h + m - j) + \sum_{x=j-m+2}^{j} c_{x-h-1,x}$$

where $c_{a,b} = 0$ for $a \leq 0$. Let $q$ be the smallest integer $p \leq q < m$, if any exists, for which $n_j \geq Q_j(q)$ for all $j$, $m \leq j \leq M$. If no such integer exists, let $q = \max(m,p)$. Then $r_0 \geq q + m$."

The operation of an SC is especially simple when it follows a nonterminal node in the code tree. Checking Figure 3.11 verifies that when any interior node is followed by the SC, the result is either one or two codewords. In the former case, the SC is simply a suffix of the single resulting codeword, while in the latter case, some of the leftmost zeros of the SC complete the string of the interior node, and the remainder of the SC (a string of the form $0^k 1^{m-q} 0^q$, where $k$ is between 0 and $q - 1$) is a valid codeword that is referred to as a reset word.

In the code tree of Figure 3.11, the reset words are 0100 and 100. When the interior node 0101 is followed by the SC, the result is 01010|0100. When 0 is followed by the SC, the result is 000100, and when 01 is followed by 00100, we obtain 0100|100. Other codewords may also have this property (i.e., they may reset certain interior nodes). This happens when a codeword is a suffix of another codeword, such as 0100, which is a suffix of 000100 in the tree of Figure 3.11.

Consider the set that consists of the SC (whose length is $m + q$ bits) and all the reset words (which are shorter). This set can reset all the nonterminal nodes, which implies a certain degree of nonuniformity in the code tree. The tree of Figure 3.11 has levels from 1 to 6, and a codeword at level $D$ is $D$ bits long. Any node (interior or not) at level $D$ must therefore have at least one codeword of length $L$ that satisfies either $D + 1 \leq L \leq D + q - 1$ (this corresponds to the case where appending the SC to the node results in two codewords) or $L = D + q + m$ (this corresponds to the case where appending the SC to the node results in one codeword). Thus, the lengths of codewords in the tree are restricted and may not truely reflect the data symbols' probabilities.

Another aspect of this restriction is that the number of codewords of the same length
that emanate from a nonterminal node is at most $2^q$. Yet another nonuniformity is that
short codewords tend to concentrate on the 0-branch of the tree. This downside of the
RHC has to be weighed against the advantage of having a resynchronizing code.

Another important factor affecting the performance of an RHC is the expected time
to resynchronize. From the way the Huffman algorithm works, we know that a codeword
of length $L$ appears in the encoded stream with probability $1/2^L$. The length of the SC
is $m+q$, so if we consider only the resynchronization provided by the SC, and ignore the
reset words, then the expected time $\tau$ to resynchronize (the inverse of the probability
of occurrence of the SC) is $2^{m+q}$. This is only an upper bound on $\tau$, because the reset
words also contribute to synchronization and reduce the expected time $\tau$. If we assume
that the SC and all the reset words reset all the nonterminal (interior) nodes, then it
can be shown that

$$\frac{1}{2^{m-1} - 2^{m+q}}$$

is a lower bound on $\tau$. Better estimates of $\tau$ can be computed for any given code tree
by going over all possible errors, determining the recovery time $\tau_i$ for each possible error
$i$, and computing the average $\tau$.

An RHC allows the decoder to resynchronize itself as quickly as possible following an
error, but there still are two points to consider: (1) the number of data symbols decoded
while the decoder is not synchronized may differ from the original number of symbols
and (2) the decoder does not know that a slippage occurred. In certain applications,
such as run-length encoding of bi-level images, where runs of identical pixels are decoded
and placed consecutively in the decoded image, these points may cause a noticeable error
when the decoded image is later examined.

A possible solution, due to [Lam and Kulkarni 96], is to take an RHC and modify
it by including an extended synchronizing codeword (ESC), which is a bit pattern that
cannot appear in any concatenation of codewords. The resulting code is no longer an
RHC, but is still an RVLC (reversible VLC, page 161). The idea is to place the ESC
at regular intervals in the encoded bitstream, say, after every 10 symbols or at the end
of each line of text or each row of pixels. When the decoder decodes an ESC, it knows
that 10 symbols should have been decoded since the previous ESC. If this is not true,
the decoder can issue an error message, warning the user of a potential synchronization
problem.

In order to understand the particular construction of an ESC, we consider all the
cases in which a codeword $c$ can be decoded incorrectly when there are errors preceding
it. Luckily, there are only two such cases, illustrated in Figure 3.12.

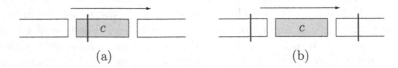

(a)                                          (b)

Figure 3.12: Two Ways to Incorrectly Decode a Codeword.

In part (a) of the figure, a prefix of $c$ is decoded as the suffix of some other codeword and a suffix of $c$ is decoded as the prefix of another codeword (in between, parts of $c$ may be decoded as several complete codewords). In part (b), codeword $c$ is concatenated with some bits preceding it and some bits following it, to form a valid codeword. This simple analysis points the way toward an ESC. It is now clear that this bit pattern should satisfy the following conditions:

1. If the ESC can be written as the concatenation of two bit strings $\alpha\beta$ where $\alpha$ is a suffix of some codeword, then $\beta$ should not be the prefix of any other codeword. Furthermore, if $\beta$ itself can be written as the concatenation of two bit strings $\gamma\delta$, then $\gamma$ can be empty or a concatenation of some codewords, and $\delta$ should be nonempty, should not be the prefix of any codeword, and no codeword should be a prefix of $\delta$. This takes care of case (a).

2. The ESC should not be a substring of any codeword. This takes care of case (b).

If the ESC satisfies these conditions, then the decoder can recognize it even in the presence of errors preceding it. Once the decoder recognizes the ESC, it reads the bits that follow until it can continue the decoding. As can be expected, the life of a researcher in the field of coding is never that simple. It may happen that an error corrupts part of the encoded bit stream into an ESC. Also, an error can corrupt the ESC pattern itself. In these cases, the decoder loses synchronization, but regains it when the next ESC is found later. It is also possible to append a fixed-size count to each ESC, such that the first ESC is followed by a count of 1, the second ESC is followed by a 2, and so on. If the latest ESC had a count $i-1$ and the next ESC has a count of $i+1$, then the decoder knows that the ESC with a count of $i$ has been missed because of such an error. If the previous ESC had a count of $i$ and the current ESC has a completely different count, then the current ESC is likely spurious.

Given a Huffman code tree, how can an ESC be constructed? We first observe that it is always possible to rearrange a Huffman code tree such that the longest codeword (the codeword of the least-probable symbol) consists of all 1's. If this codeword is $1^k$, then we construct the ESC as the bit pattern $1^k 1^{k-1} 0 = 1^{2k-1} 0$. This pattern is longer than the longest codeword, so it cannot be a substring of any codeword, thereby satisfying condition 2 above. For condition 1, consider the following. All codeword suffixes that consist of consecutive 1's can have anywhere from no 1's to $(k-1)$ 1's. Therefore, if any prefix of the ESC is the suffix of a codeword, the remainder of the ESC will be a string of the form $1^j 0$ where $j$ is between $k$ and $2k-1$, and such a string is not the prefix of any codeword, thereby satisfying condition 1.

# 3.5 Bidirectional Codes

This book, like others by the same author, is enlivened by the use of quotations and epigraphs. Today, it is easy to search the Internet and locate quotations on virtually any topic. Many web sites maintain large collections of quotations, there are many books whose full text can easily be searched, and the fast, sophisticated Internet search engines can locate many occurrences of any word or phrase.

Before the age of computers, the task of finding words, phrases, and full quotations required long visits to a library and methodical reading of many texts. For this reason, scholars sometimes devoted substantial parts of their working careers to the preparation of a concordance.

A concordance is an alphabetical list of the principal words found in a book or a body of work, with their locations (page and line numbers) and immediate contexts. Common words such as "of" and "it" are excluded. The task of manually compiling and publishing a concordance was often gargantuan, which is why concordances were generated only for highly-valued works, such as the Bible, the writings of St. Thomas Aquinas, the Icelandic sagas, and the works of Shakespeare and Wordsworth.

The emergence of the digital computer, in the late 1940s and early 1950s, has given a boost to the art of concordance compiling. Suddenly, it has become possible to store an entire body of literature in the computer and point to all the important words in it. It was also in the 1950s that Hans Peter Luhn, a computer scientist at IBM, conceived the technique of KWIC indexing. The idea was to search a concordance for a given word and display or print a condensed list of all the occurrences of the word, each with its location (page and line numbers) and immediate context. KWIC indexing remained popular for many years until it became superseded by the full text search that so many of us take for granted.

---

KWIC is an acronym for Key Word In Context, but like so many acronyms it has other meanings such as Kitchen Waste Into Compost, Kawartha World Issues Center, and Kids' Well-being Indicators Clearinghouse.

---

The following KWIC example lists the first 15 results of searching for the word **may** in a large collection (98,000 items) of excerpts from academic textbooks and introductory books located at the free concordance [amu 06].

```
No: Line:                        Concordance
 1:  73: in the National Grid - which may accompany the beginning or end of
 2:  98: of a country, however much we may analyze it into separate rules,
 3:  98: and however much the analysis may be necessary to our understanding
 4: 104: y to interpret an Act a court may sometimes alter considerably the
 5: 110: ectly acquainted with English may know the words of the statute, but
 6: 114: ime. Therefore in a sense one may speak of the Common Law as unwritten
 7: 114: e for all similar cases which may arise in the future. This binding
 8: 120:   court will not disregard. It may happen that a question has never
 9: 120: ven a higher court, though it may think a decision of a lower court
10: 122: ween the parties. The dispute may be largely a question of fact.
11: 122: nts must be delivered, and it may often be a matter of doubt how far
12: 138:  of facts. The principles, it may be, give no explicit answer to the
13: 138:  the conclusions of a science may be involved in its premisses, and
14: 144: at conception is that a thing may change and yet remain the same thing.
15: 152: ions is unmanageable, Statute may undertake the work of codification,
```

Thus, a computerized concordance that supports KWIC searches consists of a set of texts and a dictionary. The dictionary lists all the important words in the texts, each with pointers to all its occurrences. Because of their size, the texts are normally stored in compressed format where each character is assigned a variable-length code (perhaps a Huffman code). The software inputs a search term, such as **may**, from the user, locates it in the dictionary, follows the first pointer to the text, and reads the immediate context of the first occurrence of **may**. The point is that this context both precedes and follows the word, but Huffman codes (and variable-length codes in general) are designed to be read and decoded from left to right (from early to late). In general, when reading a group of variable-length codes, it is impossible to reverse direction and read the preceding code, because there is no way to tell its length. We say that variable-length codes are unidirectional, but there are applications where bidirectional (or reversible) codes are needed.

Computerized concordances and KWIC indexing is one such application. Another application, no less important, is data integrity. We already know, from the discussion of synchronous codes in Section 3.3, that errors may occur and they tend to propagate through a sequence of variable-length codes. One way to limit error propagation (slippage) is to organize a compressed file in records, where each record consists of several variable-length codes. The record is read and decoded from beginning to end, but if an error is discovered, the decoder tries to read the record from end to start. If there is only one error in the record, the two readings and decodings can sometimes be combined to isolate the error and perhaps even to correct it

There are other, much less important applications of bidirectional codes. In the early days of the digital computer, magnetic tapes were the main input/output devices. A tape is sequential storage that lends itself to linear reading and writing. If we want to read an early record from a tape, we generally have to rewind the tape and then skip forward to the desired record. The ability to read a tape backward can speed up tape input/output and make the tape look more like a random-access I/O device. Another example of the use of bidirectional codes is a little-known data structure called deque (short for double-ended queue, and sometimes spelled "dequeue"). This is a linear data structure where elements can be added to or deleted from either end. (This is in contrast to a queue, where elements can only be added to the head and deleted from the tail.) If the elements of the deque are compressed by means of variable-length codes, then bidirectional codes allow for easy access of the structure from either end.

> A good programmer is someone who always looks both ways before crossing a one-way street.
>
> —Doug Linder

It should be noted that fixed-length codes are bidirectional, but they generally do not provide any compression (the Tunstall code of Section 1.6 is an exception). We say that fixed-length codes are trivially bidirectional. Thus, we need to develop variable-length codes that are as short as possible on average but are also bidirectional. The latter requirement is important, because the bidirectional codes are going to be used for compression (where they are often called reversible VLCs or RVLCs). Anyone with even little experience in the field of variable-length codes will immediately realize that it is easy to design bidirectional codes. Simply dedicate a bit pattern $p$ to be both the prefix and suffix of all the codes, and make sure that $p$ does not appear inside a code. Thus, if $p = 101$, then 101|10011001001|101 is an example of such a code. It is easy to see that such codes can be read in either direction, but it is also obvious that they are too long, because the code bits between the suffix and prefix are restricted to patterns that do not contain $p$. A little thinking shows that $p$ doesn't even have to appear twice in each code. A code where each codeword ends with $p$ is bidirectional. A string of such codes has the form $papbpc \ldots pypz$ and it is obvious that it can be read in either direction. The taboo codes of Section 2.15 are an example of this type of variable-length code.

The average length of a code is therefore important even in the case of bidirectional codes. Recall that the most common variable-length codes can be decoded easily and uniquely because they are prefix codes and therefore instantaneous. This suggests the idea of constructing a suffix code, a code where no codeword is the suffix of another codeword. A suffix code can be read in reverse and decoded uniquely in much the same way that a prefix code is uniquely decoded. Thus, a code that is both a prefix code and a suffix code is bidirectional. Such a code is termed *affix* (although some authors use the terms "biprefix" and "never-self-synchronizing").

The next few paragraphs are based on [Fraenkel and Klein 90], a work that analyses Huffman codes, looking for ways to construct affix Huffman codes or modify existing Huffman codes to make them affix and thus bidirectional. The authors first show how to start with a given affix Huffman code $C$ and double its size. The idea is to take every codeword $c_i$ in $C$ and create two new codewords from it by appending a bit to it. The two new codewords are therefore $c_i0$ and $c_i1$. The resulting set of codewords is affix as can be seen from the following argument.

1. No other codeword in $C$ starts with $c_i$ ($c_i$ is not the prefix of any other codeword). Therefore, no other codeword in the new code starts with $c_i$. As a result, no other codeword in the new code starts with $c_i0$ or $c_i1$.

2. Similarly, $c_i$ is not the suffix of any other codeword in $C$, therefore neither $c_i0$ nor $c_i1$ are suffixes of a codeword in the new code.

Given the affix Huffman code 01, 000, 100, 110, 111, 0010, 0011, 1010, and 1011, we apply this method to double it and construct the code 010, 0000, 1000, 1100, 1110, 00100, 00110, 10100, 011, 0001, 1001, 1101, 1111, 00101, 00111, and 10101. A simple check verifies that the new code is affix. The conclusion is that there are infinitely many affix Huffman codes.

On the other hand, there are cases where affix Huffman codes do not exist. Consider, for example, codewords of length 1. In the trivial case where there are only two codewords, each is a single bit. This case is trivial and is also a fixed-length code. Thus, a variable-length code can have at most one codeword of length 1 (a single bit). If a code has such a codeword, then it is the suffix of other codewords, because in a complete prefix code, codewords must end with both 0 and 1. Thus, a code one of whose codewords is of length 1 cannot be affix. Such Huffman codes exist for sets of symbols with skewed probabilities. In fact, it is known that the existence of a codeword $c_i$ of length 1 in a Huffman code implies that the probability of the symbol that corresponds to $c_i$ must be greater than 1/3.

The authors then describe a complex algorithm (not listed here) to construct affix Huffman codes for cases where such codes exist.

There are many other ways to construct RVLCs from Huffman codes. Table 3.13 (after [Laković and Villasenor 03], see also Table 3.22) lists a set of Huffman codes for the 26 letters of the English alphabet together with three RVLC codes for the same symbols. These codes were constructed by algorithms proposed by [Takishima et al. 95], [Tsai and Wu 01a], and [Laković and Villasenor 03].

The remainder of this section describes the extension of Rice codes and exponential Golomb (EG) codes to bidirectional codes (RVLCs). The resulting bidirectional codes have the same average length as the original, unidirectional Rice and EG codes. They have been adopted by the International Telecommunications Union (ITU) for use

| $p$ | | Huffman | Takishima | Tsai | Laković |
|---|---|---|---|---|---|
| E | 0.14878 | 001 | 001 | 000 | 000 |
| T | 0.09351 | 110 | 110 | 111 | 001 |
| A | 0.08833 | 0000 | 0000 | 0101 | 0100 |
| O | 0.07245 | 0100 | 0100 | 1010 | 0101 |
| R | 0.06872 | 0110 | 1000 | 0010 | 0110 |
| N | 0.06498 | 1000 | 1010 | 1101 | 1010 |
| H | 0.05831 | 1010 | 0101 | 0100 | 1011 |
| I | 0.05644 | 1110 | 11100 | 1011 | 1100 |
| S | 0.05537 | 0101 | 01100 | 0110 | 1101 |
| D | 0.04376 | 00010 | 00010 | 11001 | 01110 |
| L | 0.04124 | 10110 | 10010 | 10011 | 01111 |
| U | 0.02762 | 10010 | 01111 | 01110 | 10010 |
| P | 0.02575 | 11110 | 10111 | 10001 | 10011 |
| F | 0.02455 | 01111 | 11111 | 001100 | 11110 |
| M | 0.02361 | 10111 | 111101 | 011110 | 11111 |
| C | 0.02081 | 11111 | 101101 | 100001 | 100010 |
| W | 0.01868 | 000111 | 000111 | 1001001 | 100011 |
| G | 0.01521 | 011100 | 011101 | 0011100 | 1000010 |
| Y | 0.01521 | 100110 | 100111 | 1100011 | 1000011 |
| B | 0.01267 | 011101 | 1001101 | 0111110 | 1110111 |
| V | 0.01160 | 100111 | 01110011 | 1000001 | 10000010 |
| K | 0.00867 | 0001100 | 00011011 | 00111100 | 10000011 |
| X | 0.00146 | 00011011 | 000110011 | 11000011 | 11100111 |
| J | 0.00080 | 000110101 | 0001101011 | 100101001 | 100000010 |
| Q | 0.00080 | 0001101001 | 00011010011 | 0011101001 | 1000000010 |
| Z | 0.00053 | 0001101000 | 000110100011 | 1001011100 | 1000000111 |
| Avg. length | | 4.15572 | 4.36068 | 4.30678 | 4.25145 |

Table 3.13: Huffman and Three RVLC Codes for the English Alphabet.

in the video coding parts of MPEG-4, and especially in the H.263v2 (also known as H.263+ or H.263 1998) and H263v3 (also known as H.263++ or H.263 2000) video compression standards [T-REC-h 06]. The material presented here is based on [Wen and Villasenor 98].

The Rice codes (Section 2.24) are a special case of the more general Golomb code, where the parameter $m$ is a power of 2 ($m = 2^k$). Once the base $k$ has been chosen, the Rice code of the unsigned integer $n$ is constructed in two steps: (1) Separate the $k$ least-significant bits (LSBs) of $n$. They become the LSBs of the Rice code. (2) Code the remaining $j = \lfloor n/2^k \rfloor$ bits in unary as either $j$ zeros followed by a 1 or $j$ 1's followed by a 0. This becomes the most-significant part of the Rice code. This code is therefore easily constructed with a few logical operations.

Decoding is also simple and requires only the value of $k$. The decoder scans the most-significant 1's until it reaches the first 0. This gives it the value of the most-significant part of $n$. The least-significant part of $n$ is the $k$ bits following the first 0. This simple decoding points the way to designing a bidirectional Rice code. The second part is always $k$ bits long, so it can be read in either direction. To also make the first (unary) part bidirectional, we change it from $111\ldots10$ to $100\ldots01$, unless it is a single bit, in which case it becomes a single 0. Table 3.14 lists several original and bidirectional Rice codes. It should be compared with Table 2.50.

|   |   |   | No. of | Rice | Rev. |
|---|---|---|---|---|---|
| $n$ | Binary | LSB | 1's | Code | Rice |
| 0 | 0 | 00 | 0 | 0\|00 | 0\|00 |
| 1 | 1 | 01 | 0 | 0\|01 | 0\|01 |
| 2 | 10 | 10 | 0 | 0\|10 | 0\|10 |
| 3 | 11 | 11 | 0 | 0\|11 | 0\|11 |
| 4 | 100 | 00 | 1 | 10\|00 | 11\|00 |
| 5 | 101 | 01 | 1 | 10\|01 | 11\|01 |
| 6 | 110 | 10 | 1 | 10\|10 | 11\|10 |
| 7 | 111 | 11 | 1 | 10\|11 | 11\|11 |
| 8 | 1000 | 00 | 2 | 110\|00 | 101\|00 |
| 11 | 1011 | 11 | 2 | 110\|11 | 101\|11 |
| 12 | 1100 | 00 | 3 | 1110\|00 | 1001\|00 |
| 15 | 1111 | 11 | 3 | 1110\|11 | 1001\|11 |

Table 3.14: Original and Bidirectional Rice Codes.

An interesting property of the Rice codes is that there are $2^k$ codes of each length and the lengths start at $k+1$ (the prefix is at least one bit, and there is a $k$-bit suffix). Thus, for $k = 3$ there are eight codes of length 4, eight codes of length 5, and so on. For certain probability distributions, we may want the number of codewords of length $L$ to grow exponentially with $L$, and this feature is offered by a parametrized family of codes known as the exponential Golomb codes. These codes were first proposed in [Teuhola 78] and are also identical to the triplet $(s, 1, \infty)$ of start-step-stop codes. They perform well for probability distributions that are exponential but are taller than average and have a wide tail.

The exponential Golomb codes depend on the choice of a nonnegative integer parameter $s$ (that becomes the length of the suffix of the codewords). The nonnegative integer $n$ is encoded in the following steps:

1. Compute $w = 1 + \lfloor n/2^s \rfloor$.
2. Compute $f_{2^s}(n) = \lfloor \log_2 \left[ 1 + \frac{n}{2^s} \right] \rfloor$. This is the number of bits following the leftmost 1 in the binary representation of $w$.
3. Construct the codeword $\text{EG}(n)$ as the unary representation of $f_{2^s}(n)$, followed by the $f_{2^s}(n)$ least-significant bits in the binary representation of $w$, followed by the $s$ least-significant bits in the binary representation of $n$.

Thus, the length of this codeword is the sum

$$l(n) = 1 + 2f_{2^s}(n) + s = 1 + 2 \left\lfloor \log_2 \left[ 1 + \frac{n}{2^s} \right] \right\rfloor + s = P + s,$$

where $P$ is the prefix (whose length is always odd) and $s$ is the suffix of a codeword. Because the logarithm is truncated, the length increases by 2 each time the logarithm increases by 1, i.e., each time $1 + n/2^s$ is a power of 2.

(As a side note, it should be mentioned that the exponential Golomb code can be further generalized by substituting an arbitrary positive integer parameter $m$ for the expression $2^s$. Such codes can be called generalized exponential Golomb codes.)

As an example, we select $s = 1$ and determine the exponential Golomb code of $n = 11_{10} = 1011_2$. We compute $w = 1 + \lfloor 11/2^s \rfloor = 6 = 110_2$, $f_2(11) = \lfloor \log_2(11/2) \rfloor = 2$, and construct the three parts $110|10|1$.

Table 3.15 lists (in column 2) several examples of the exponential Golomb codes for $s = 1$. Each code has an $s$-bit suffix and the prefixes get longer with $n$. The table also illustrates (in column 3) how these codes can be modified to become bidirectional. The idea is to have the prefix start and end with 1's, to fill the odd-numbered bit positions of the prefix (except the two extreme ones) with zeros, and to fill the even-numbered positions with bit patterns that represent increasing integers. Thus, for $s = 1$, the 16 $(7+1)$-bit codewords (i.e., the codewords for $n = 14$ through $n = 29$) have a 7-bit prefix of the form $1x0y0z1$ where the bits $xyz$ take the eight values 000 through 111. Pairs of consecutive codewords have the same prefix and differ in their 1-bit suffixes.

|   | Exp. | Rev. |
|---|---|---|
| $n$ | Golomb | exp. |
| 0 | 0\|0 | 0\|0 |
| 1 | 0\|1 | 0\|1 |
| 2 | 100\|0 | 101\|0 |
| 3 | 100\|1 | 101\|1 |
| 4 | 101\|0 | 111\|0 |
| 5 | 101\|1 | 111\|1 |
| 6 | 11000\|0 | 10001\|0 |
| 7 | 11000\|1 | 10001\|1 |
| 8 | 11001\|0 | 10011\|0 |
| 9 | 11001\|1 | 10011\|1 |
| 10 | 11010\|0 | 11001\|0 |
| 11 | 11010\|1 | 11001\|1 |

Table 3.15: Original and Bidirectional Exponential Golomb Codes.

Here is how the decoder can read such codewords in reverse and identify them. The decoder knows the value of $s$, so it first reads the $s$-bit suffix. If the next bit (the rightmost bit of the prefix) is 0, then the entire prefix is this single bit and the suffix determines the value of $n$ (between 0 and $2^s - 1$). Otherwise, the decoder reads bits from right to left, isolating the bits with even indexes (shown in boldface in Table 3.15) and concatenating them from right to left, until an odd-indexed bit of 1 is found. The total number of bits read is the length $P$ of the prefix. All the prefixes of length $P$ differ only in their even-numbered bits, and a $P$-bit prefix (where $P$ is always odd) has $(P - 1)/2$ such bits. Thus, there are $2^{(P-1)/2}$ groups of prefixes, each with $2^s$ identical prefixes. The decoder identifies the particular $n$ in a group by the $s$-bit suffix.

The bidirectional exponential Golomb codes therefore differ from the (original) exponential Golomb codes in their construction, but the two types have the same lengths.

**The magical exclusive-OR.** The methods presented earlier are based on sets of Huffman, Rice, and exponential Golomb codes that are modified and restricted in order to become bidirectional. In contrast, the next few paragraphs present a method, due to [Girod 99], where any set of prefix codes $B_i$ can be transformed to a bitstring $C$ that can be decoded in either direction. The method is simple and requires only logical operations and string reversals. It is based on a well-known "magical" property of the exclusive-OR (XOR) logical operation, and its only downside is the addition of a few extra zero bits. First, a few words about the XOR operation and what makes it special.

The logical OR operation is familiar to many. It receives two bits $a$ and $b$ as its inputs and it outputs one bit. The output is 1 if $a$ or $b$ or both are 1's. The XOR operation is similar but it excludes the case where both inputs are 1's. Thus, the result of (1 XOR 1) is 0. Both logical operations are summarized in the table.

|          |      |
|---------:|------|
| $a$      | 0011 |
| $b$      | 0101 |
| $a$ OR $b$  | 0111 |
| $a$ XOR $b$ | 0110 |

The table also shows that the XOR of any bit $a$ with 0 is $a$.

What makes the XOR so useful is the following property: if $c = a$ XOR $b$, then $b = a$ XOR $c$ and $a = b$ XOR $c$. This property is easily verified and it has made the XOR very popular in many applications (see page 357 of [Salomon 03] for an interesting example).

This useful property of the XOR is now exploited as follows. The first idea is to start with a string of $n$ data symbols and encode them with a prefix code to produce a string $B$ of $n$ codewords $B_1 B_2 \ldots B_n$. Now reverse each codeword $B_i$ to become $B_i'$ and construct the string $B' = B_1' B_2' \ldots B_n'$. Next, compute the final result $C = B$ XOR $B'$. The hope is that the useful property of the XOR will enable us to decode $B$ from $C$ with the relation $B = C$ XOR $B'$. This simple scheme does not work because the relation requires knowledge of $B'$. We don't know string $B'$, but we know that its components are closely related to those of string $B$. Thus, the following trick becomes the key to this elegant method. Denote by $L$ the length of the longest codeword $B_i$, append $L$ zeros to $B$ and prepend $L$ zeros to $B'$ (more than $L$ zeros can be used, but not fewer). Now perform the operation $C = (B \underbrace{00\ldots0}_{L})$ XOR $(\underbrace{00\ldots0}_{L} B')$. It is clear that the first $L$ bits of $C$ are identical to the first $L$ bits of $B$. Because of the choice of $L$, those $L$ bits constitute at least the first codeword $B_1$ (there may be more than one codeword and there may also be a remainder), so we can immediately reverse it to obtain $B_1'$. Now we can read more bits from $C$ and XOR them with $B_1'$ to obtain more bits with parts of codewords from $B$. This unusual decoding procedure is best illustrated by an example.

We start with five symbols $a_1$ through $a_5$ and assign them the variable-length prefix codes 0, 10, 111, 1101, and 1100. The string of symbols $a_2 a_1 a_3 a_5 a_4$ is compressed by this code to the string $B = 10|0|111|1100|1101$. Reversing each codeword produces $B' = 01|0|111|0011|1011$. The longest codeword is four bits long, so we select $L = 4$.

We append four zeros to $B$ and prepend four zeros to $B'$. Exclusive-ORing the strings yields

$$B = 1001\ 1111\ 0011\ 010000$$
$$B' = 0000\ 0101\ 1100\ 111011$$
$$C = 1001\ 1010\ 1111\ 101011$$

Decoding is done in the following steps:

1. XOR the first $L$ bits of $C$ and $B'$. This results in $1001 \oplus 0000 = 1001 \to a_2 a_1 1$. This step decodes the first two symbols and leaves a "remainder" of 1 for the next step.

2. A total of three bits were decoded in the previous step, so the current step XORs the next three bits of $C$ and $B'$. This results in $101 \oplus 010 = 111$. Prepending the remainder from the previous step yields 1111, which is decoded to $a_3$ and a remainder of 1.

3. A total of three bits were decoded in the previous step, so the current step XORs the next three bits of $C$ and $B'$, which results in $011 \oplus 111 = 100$. Prepending the remainder from the previous step yields 1100 which is decoded to $a_5$ with no remainder.

4. Four bits were decoded in the previous step, so the current step XORs the next four bits of $C$ and $B'$, which results in $1110 \oplus 0011 = 1101$, which is decoded to $a_4$.

5. Four bits were decoded in the previous step, so the current step XORs the next four bits of $C$ and $B'$, which results in $1011 \oplus 1011 = 0000$. This indicates the successful end of the decoding procedure.

If any data becomes corrupted, the last step will produce something other than $L$ zeros, indicating unsuccessful decoding.

Decoding in the reverse direction is identical, except that $C$ is fed to the decoder from end to start (from right to left) to produce substrings of $B$, which are then decoded and also reversed (to become codewords of $B'$) and sent to the XOR. The first step XORs the reverse of the last four bits of $C$ with the the reverse of the last four bits of $B$ (the four zeros) to produce 1101, which is decoded as $a_4$, reversed, and sent to the XOR.

The only overhead is the extra $L$ bits, but $L$ is normally a small number. Figure 3.16 shows the encoder and decoder of this method.

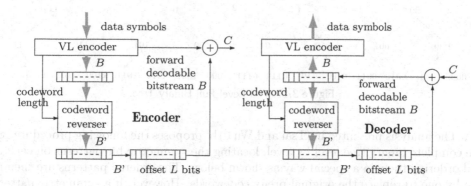

Figure 3.16: XOR-Based Encoding and Decoding.

# 3.6 Symmetric Codes

A symmetric code is one where every codeword is symmetric. Such a codeword looks the same when read in either direction, which is why symmetric codes are a special case of reversible codes. We can expect a symmetric code to feature higher average length compared with other codes, because the requirement of symmetry restricts the number of available bit patterns of any given length.

The material presented here describes one way of selecting a set of symmetric codewords. It is based on [Tsai and Wu 01b], which is an extension of [Takishima et al. 95]. The method starts from a variable-length prefix code, a set of prefix codewords of various lengths, and replaces the codewords with symmetric bit patterns that have the same or similar lengths and also satisfy the prefix property. Figure 3.17 is a good starting point. It shows a complete binary tree with four levels. The symmetric bit patterns at each level are underlined (as an aside, it can be proved by induction that there are $2^{\lfloor (i+1)/2 \rfloor}$ such patterns on level $i$) and it is clear that, even though the number of symmetric patterns is limited, every path from the root to a leaf passes through several such patterns. It is also clear from the figure that any bit pattern is the prefix of all the patterns below it on the same path. Thus, for example, selecting the pattern 00 implies that we cannot later select any of the symmetric patterns 000, 0000, or any other symmetric pattern below them on the same path. Selecting the 3-bit symmetric pattern 010, on the other hand, does not restrict the choice of 4-bit symmetric patterns. (It will restrict the choice of longer patterns, such as 01010 or 010010, but we are more interested in short patterns.) Thus, a clever algorithm is needed, to select those symmetric patterns at any level that will maximize the number of available symmetric patterns on the levels immediately below them.

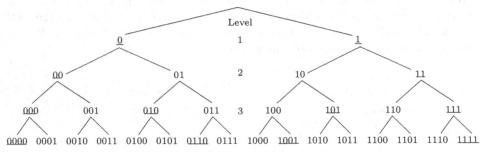

Figure 3.17: A 4-Level Full Binary Tree.

The analysis presented in [Tsai and Wu 01b] proposes the following procedure. Scan the complete binary tree level by level, locating the symmetric bit patterns on each level and ordering them in a special way as shown below. The ordered patterns are then used one by one to replace the original prefix codewords. However, if a symmetric pattern $P$ violates the prefix property (i.e., if a previously-assigned pattern is a prefix of $P$), then $P$ should be dropped.

To order the symmetric patterns of a level, go through the following steps:

1. Ignore the leftmost bit of the pattern.

2. Examine the remaining bits and determine the maximum number $M$ of least-significant bits that are still symmetric (the maximum number of symmetric bit suffixes).

Table 3.18 lists the orders of the symmetric patterns on levels 3 through 6 of the complete binary tree. In this table, $M$ stands for the maximum number of symmetric bit suffixes and CW is a symmetric pattern. For example, the 5-bit pattern 01110 has $M = 1$ because when we ignore its leftmost bit, the remaining four bits 1110 are asymmetric (only the rightmost bit is symmetric). On the other hand, pattern 10101 has $M = 3$ because the three rightmost bits of 0101 are symmetric.

The symmetric bit patterns on each level should be selected and assigned in the order shown in the table (order of increasing $M$). Thus, if we need 3-bit symmetric patterns, we should first select 010 and 101, and only then 000 and 111. Selecting them in this order maximizes the number of available symmetric patterns on levels 4 and 5.

| Level 3 | | Level 4 | | Level 5 | | Level 6 | |
|---|---|---|---|---|---|---|---|
| $M$ | CW | $M$ | CW | $M$ | CW | $M$ | CW |
| 1 | 010 | 1 | 0110 | 1 | 01110 | 1 | 011110 |
| 1 | 101 | 1 | 1001 | 1 | 10001 | 1 | 100001 |
| 2 | 000 | 3 | 0000 | 2 | 00100 | 2 | 001100 |
| 2 | 111 | 3 | 1111 | 2 | 11011 | 2 | 110011 |
| | | | | 3 | 01010 | 3 | 010010 |
| | | | | 3 | 10101 | 3 | 101101 |
| | | | | 4 | 00000 | 5 | 000000 |
| | | | | 4 | 11111 | 5 | 111111 |

Table 3.18: Symmetric Codewords on Levels 3 Through 6, Ordered by Symmetric Bit Suffixes.

The following, incomplete, example illustrates the operation of this interesting algorithm. Figure 3.19 shows a Huffman code for the 26 letters of the English alphabet (the individual probabilities of the letters appear in Figure 1.19). The first Huffman codeword is 000. It happens to be symmetric, but Table 3.18 tells us to replace it with 010. It is immediately clear that this is a good choice (as would have been the choice of 101), because Figure 3.17 shows that the path that leads down from 010 to level 4 passes through patterns 0100 and 0101 which are not symmetric. Thus, the choice of 010 does not restrict the future choice of 4-bit symmetric codewords.

Figure 3.19 requires six 4-bit symmetric codewords, but there are only four such patterns. They are 0110, 1001, 0000, and 1111, selected in increasing number of symmetric bit suffixes. Pattern 010 is not a prefix of any of them, so they are assigned as the new, symmetric codewords of T, A, O, and N. The remaining two symmetric codewords (of R and I) on this level will have to be longer (five bits each). Figure 3.19 requires 14 5-bit symmetric codewords (and two more, for R and I, are still needed), but there are only eight symmetric 5-bit patterns. Two of them cannot be used because of prefix violation. Pattern 010 is a prefix of 01010 and pattern 0000 is a prefix of 00000. The remaining six 5-bit codewords of Table 3.18 are selected and assigned to the letters R through L, while the letters C through V will have to be assigned longer symmetric codewords.

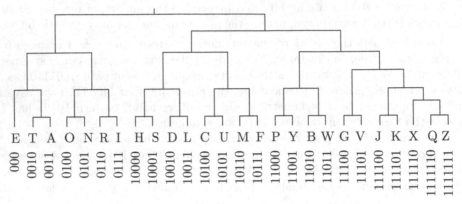

Figure 3.19: A Huffman Code for the 26-Letter Alphabet.

It seems that the symmetric codewords are longer on average than the original Huffman codewords, and this is true in general, because the number of available symmetric codewords of any given length is limited. However, as we go down the tree to longer and longer codewords, the situation improves, because every other level in the complete binary tree doubles the number of available symmetric codewords. Reference [Tsai and Wu 01b] lists Huffman codes and symmetric codes for the 26 letters with average lengths of 4.156 and 4.61 bits per letter, respectively.

The special ordering rules above seem arbitrary, but are not difficult to understand intuitively. As an example, we examine the eight symmetric 6-bit patterns. The two patterns of $M = 5$ become prefixes of patterns at level 7 simply by appending a 0 or a 1. These patterns should therefore be the last ones to be used because they restrict the number of available 7-bit symmetric patterns. The two patterns for $M = 3$ are better, because they are not the prefixes of any 7-bit or 8-bit symmetric patterns. It is obvious that appending one or two bits to 010010 or 101101 cannot result in a symmetric pattern. Thus, these two patterns are prefixes of 9-bit (or longer) symmetric patterns. When we examine the two patterns for $M = 2$, the situation is even better. Appending one, two, or even three bits to 100001 or 001100 cannot result in a symmetric pattern. These patterns are the prefixes of 10-bit (or longer) patterns. Similarly, the two 6-bit patterns for $M = 1$ are the best, because their large asymmetry implies that they can only be prefixes of 11-bit (or longer) patterns. Selecting these two patterns does not restrict the number of symmetric 7-, 8-, 9-, or 10-bit patterns.

# 3.7 VLEC Codes

The simplest approach to robust variable-length codes is to add a parity bit to each codeword. This approach combines compression (source coding) and reliability (channel coding), but keeps the two separate. The two aims are contradictory because the former removes redundancy while the latter adds redundancy. However, any approach to robust compressed data must necessarily reduce compression efficiency. This simple approach has an obvious advantage; it makes it easy to respond to statistical changes in both the source and the channel. If it turns out that the channel is noisy, several parity bits may be appended to each codeword, but the codewords themselves do not have to be modified. If it turns out that certain data symbols occur with higher probability than originally thought, their codewords can be replaced with shorter ones without having to modify the error-control scheme.

A different approach to the same problem is to construct a set of variable-length codewords that are sufficiently distant from one another (distant in the sense of Hamming distance). A set of such codewords can be called a variable-length error-correcting (VLEC) code. The downside of this approach is that any changes in source or channel statistics requires a new set of codewords, but the hope is that this approach may result in better overall compression. This approach may be justified if it meets the following two goals:

1. The average length of the codewords turns out to be shorter than the average length of the codewords with parity bits of the previous approach.

2. A decoding algorithm is found that can exploit the large distance between variable-length codewords to detect and even correct errors. Even better, such an algorithm should be able to recover synchronization in cases where bits get corrupted in the communications channel.

Several attempts to develop such codes are described here.

**Alpha-Prompt Codes.** The main idea of this technique is to associate each codeword $c$ with a set $\alpha(c)$ of bit patterns of the same length that are at certain Hamming distances from $c$. If codeword $c$ is transmitted and gets corrupted to $c'$, the decoder checks the entire set $\alpha(c)$ and selects the pattern nearest $c'$. The problem is that the decoder doesn't know the length of the next codeword, which is why a practical method based on this technique should construct the sets $\alpha(c)$ in a special way.

We assume that there are $s_1$ codewords of length $L_1$, $s_2$ codewords of length $L_2$, and so on, up to length $m$. The set of all the codewords $c$ and all the bit patterns in the individual sets $\alpha(c)$ is constructed as a prefix code (i.e., it satisfies the prefix property and is therefore instantaneous). The decoder starts by reading the next $L_1$ bits from the input. Let's denote this value by $t$. If $t$ is a valid codeword $c$ in $s_1$, then $t$ is decoded to $c$. Otherwise, the decoder checks all the leftmost $L_1$ bits of all the patterns that are $L_1$ bits or longer and compares $t$ to each of them. If $t$ is identical to a pattern in set $\alpha(c_i)$, then $t$ is decoded to $c_i$. Otherwise, the decoder again goes over the leftmost $L_1$ bits of all the patterns that are $L_1$ bits or longer and selects the one whose distance from $t$ is minimal. If that pattern is a valid codeword, then $t$ is decoded to it. Otherwise, the decoder reads a few more bits, for a total of $L_2$ bits, and repeats this process for the first $L_2$ bits of all the patterns that are $L_2$ bits long or longer.

This complex algorithm (proposed by [Buttigieg 95]) always decodes the input to some codeword if the set of all the codewords $c$ and all the bit patterns in the individual sets $\alpha(c)$ is a prefix code.

**VLEC Tree.** We assume that our code has $s_1$ codewords of length $L_1$, $s_2$ codewords of length $L_2$, and so on, up to length $m$. The total number of codewords $c_i$ is $s$. As an example, consider the code $a = 000$, $b = 0110$, and $c = 1011$ (one 3-bit and two 4-bit codewords).

The VLEC tree method attempts to correct errors by looking at an entire transmission and mapping it to a special VLEC tree. Each node in this tree is connected to $s$ other nodes on lower levels with edges $s$ that are labeled with the codewords $c_i$. Figure 3.20 illustrates an example with the code above. Each path of $e$ edges from the root to level $b$ is therefore associated with a different string of $e$ codewords whose total length is $b$ bits. Thus, node [1] in the figure corresponds to string $cba = 1011|0110|000$ (11 bits) and node [2] corresponds to the 12-bit string $bcc = 0110|1011|1011$. There are many possible strings of codewords, but it is also clear that the tree grows exponentially and quickly becomes very wide (the tree in the figure corresponds to strings of up to 12 bits).

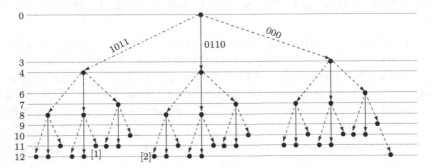

Figure 3.20: A VLEC Tree.

Assuming that the only transmission errors are corrupted bits (i.e., no bits are inserted to or deleted from an encoded string by noise in the communications channel), if an encoded string of $b$ bits is sent, the decoder will receive $b$ bits, some possibly bad. Any $b$-bit string corresponds to a node on level $b$, so error-correcting is reduced to the problem of selecting one of the paths that end on that level. The obvious solution is to measure the Hamming distances between the received $b$-bit string and all the paths that end on level $b$ and select the path with the minimum distance. This simple solution, however, is impractical for the following reasons:

■   Data symbols (and therefore their codewords) have different probabilities, which is why not all $b$-bit paths are equally likely. If symbol $a$ is very probable, then paths with many occurrences of $a$ are more likely and should be given more weight. Assume that symbol $a$ in our example has probability 0.5, while symbols $b$ and $c$ occur with probability 0.25 each. A probable path such as $aabaca$ should be assigned the weight $4 \times 0.5 + 2 \times 0.25 = 2.5$, while a less probable path, such as $bbccca$ should be assigned the smaller weight $5 \times 0.25 + 0.5 = 1.75$. When the decoder receives a $b$-bit string $S$, it should (1) compute the Hamming distance between $S$ and all the $b$-bit paths $P$, (2) divide each distance by the weight of the path, and (3) select the path with the smallest weighted distance. Thus, if two paths with weights 2.5 and 1.75 are at a Hamming distance of 2 from $S$, the decoder computes $2/2.5 = 0.8$ and $2/1.75 = 1.14$ and selects the former path.

■   A VLEC tree for even a modest-size alphabet grows too wide very quickly and becomes unmanageable for even short messages. Thus, an algorithm is needed for the decoder to decode an encoded message in short segments. In cases where many errors are rare, a possible approach may be to read the encoded string and encode it normally until the decoder loses synchronization (say, at point $A$). The decoder then skips bits until it regains synchronization (at point $B$), and then emplo... the VLEC ... to the input from $A$ to $B$.

**VLEC Trellis.** The second point above, of the tree getting too big very quickly, can be overcome by replacing the tree with a trellis structure, as shown in Figure 3.21.

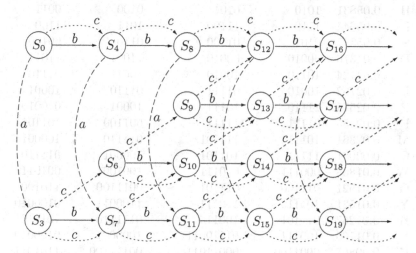

Figure 3.21: A VLEC Trellis.

The trellis is constructed as follows:

1. Start with state $S_0$.

2. From each state, construct $m$ edges to other states, where each edge corresponds to one of the $m$ codeword lengths $L_i$. Thus, state $S_i$ will be connected to states $S_{i+L_1}$, $S_{i+L_2}$, up to $S_{i+L_m}$. If any of these states does not exist, it is created when needed.

> A trellis is a structure, usually made from interwoven pieces of wood, bamboo or metal that supports many types of climbing plant such as sweet peas, grapevines and ivy. —wikipedia.com

Notice that the states are arranged in Figure 3.21 in stages, where each stage is a vertical set of states. A comparison of Figures 3.20 and 3.21 shows their similarities, but also shows the important difference between them. Beyond a certain point (in our example, beyond the third stage), the trellis stages stop growing in size regardless of the length of the trellis. It is this difference that makes it possible to develop a reliable decoding algorithm (a modified Viterbi algorithm, described on page 69 of [Buttigieg 95], but not discussed here). An important feature of this algorithm is that it employs a metric (a weight assigned to each trellis edge), and it is possible to define this metric in a way that takes into account the different probabilities of the trellis edges.

# 3.7.1 Constructing VLEC Codes

| | $p$ | Huffman | Takishima | Tsai | Laković |
|---|---|---|---|---|---|
| E | 0.14878 | 001 | 001 | 000 | 000 |
| T | 0.09351 | 110 | 110 | 111 | 111 |
| A | 0.08833 | 0000 | 0000 | 0101 | 0101 |
| O | 0.07245 | 0100 | 0100 | 1010 | 1010 |
| R | 0.06872 | 0110 | 1000 | 0010 | 0110 |
| N | 0.06498 | 1000 | 1010 | 1101 | 1001 |
| H | 0.05831 | 1010 | 0101 | 0100 | 0011 |
| I | 0.05644 | 1110 | 11100 | 1011 | 1100 |
| S | 0.05537 | 0101 | 01100 | 0110 | 00100 |
| D | 0.04376 | 00010 | 00010 | 11001 | 11011 |
| L | 0.04124 | 10110 | 10010 | 10011 | 01110 |
| U | 0.02762 | 10010 | 01111 | 01110 | 10001 |
| P | 0.02575 | 11110 | 10111 | 10001 | 010010 |
| F | 0.02455 | 01111 | 11111 | 001100 | 101101 |
| M | 0.02361 | 10111 | 111101 | 011110 | 100001 |
| C | 0.02081 | 11111 | 101101 | 100001 | 011110 |
| W | 0.01868 | 000111 | 000111 | 1001001 | 001011 |
| G | 0.01521 | 011100 | 011101 | 0011100 | 110100 |
| Y | 0.01521 | 100110 | 100111 | 1100011 | 0100010 |
| B | 0.01267 | 011101 | 1001101 | 0111110 | 1011101 |
| V | 0.01160 | 100111 | 01110011 | 1000001 | 0100010 |
| K | 0.00867 | 0001100 | 00011011 | 00111100 | 1101011 |
| X | 0.00146 | 00011011 | 000110011 | 11000011 | 10111101 |
| J | 0.00080 | 000110101 | 0001101011 | 100101001 | 010000010 |
| Q | 0.00080 | 0001101001 | 00011010011 | 0011101001 | 0100000010 |
| Z | 0.00053 | 0001101000 | 000110100011 | 1001011100 | 1011111101 |
| Avg.  length | | 4.15572 | 4.36068 | 4.30678 | 4.34534 |

Table 3.22: Huffman and Three RVLC Codes for the English Alphabet.

How can we construct an RVLC with a free distance greater than 1? One such algorithm, proposed by [Laković and Villasenor 02] is an extension of an older algorithm due to [Tsai and Wu 01a], which itself is based on the work of [Takishima et al. 95]. This algorithm starts from a set of Huffman codes. It then goes over these codewords level by level, from short to long codewords and replaces some of them with patterns taken from a complete binary tree, similar to what is described in Section 3.6, making sure that the prefix property is not violated while also ascertaining that the free distance of all the codewords that have so far been examined does not drop below the target free distance. If a certain pattern from the binary tree results in a free distance that is too small, that pattern is skipped.

Table 3.22 (after [Laković and Villasenor 02], see also the very similar Table 3.13) lists a set of Huffman codes for the 26 letters of the English alphabet together with three RVLC codes for the same symbols. The last of these codes has a free distance of 2. These codes were constructed by algorithms proposed by [Takishima et al. 95], [Tsai and Wu 01a], and [Laković and Villasenor 02].

**Summary**. Errors are a fact of life, which is why any practical method for coding, storing, or transmitting data should consider the use of robust codes. The various variable-length codes described in this chapter are robust, which makes them ideal choices for applications where both data compression and data reliability are needed.

> Good programmers naturally write neat code when left
> to their own devices. But they also have an array of battle
> tactics to help write robust code on the front line.
> —Pete Goodliffe, *Code Craft: The Practice of Writing Excellent Code*

# Summary and Unification

The wide experience currently available with variable-length codes indicates that it is easy to develop good variable-length codes for large integers. The Elias omega, the ternary comma, and the punctured codes all do well in compressing large integers and there is not much difference between them.

When the data is dominated by many small values, none of the variable-length codes performs much better than others (although the gamma and omega codes seem to perform a bit better in such cases). It is difficult (perhaps even impossible) to develop a variable-length code that will consistently encode small integers with short codes.

The previous two paragraphs suggest that the omega code does well with both small and large data values, but it doesn't do as well in the intermediate range, where it introduces new prefix elements (length groups). Thus, even this code is not ideal for arbitrary data distributions.

**Unification.** Today (early 2007), after several decades of intensive research in the area of variable-length codes, many codes have been conceived, developed, analyzed, and applied in practical situations to compress various types of data. The main classes of variable-length codes are listed in the Introduction, and it is obvious that they are based on different approaches and different principles, and possess different properties. Nevertheless, the human tendency to unify disparate entities is strong, and the next few paragraphs describe an attempt to unify variable-length codes.

Some of the codes presented here constitute families that depend on a parameter (or a small number of parameters). For each choice of the parameter, the family reduces to a code whose members are codewords. Following is a list of these codes.

The Stout codes of Section 2.11, the Golomb code of Section 2.23, the Rice codes (Section 2.24), the exponential Golomb codes (proposed by [Teuhola 78]), the start-step-stop codes of Fiala and Greene (Section 2.2), the start/stop codes of Pigeon (Section 2.3), the taboo codes of Section 2.15, the Capocelli code of Section 2.20.1, and the subexponential code of Howard and Vitter (Section 2.25).

The fact that many variable-length codes depend on a parameter and consist of families suggests a way to unify them. The idea described here is due to Paul Howard who, in [Howard 98], termed such codes unary prefix/suffix, or UP/S. The codewords of a UP/S code are organized in sets $S_i$ such that the codewords of the first few integers

0, 1, 2 ... are members of set $S_0$, the following codewords are members of set $S_1$, and so on. Each set is completely filled up before the next set starts. A codeword consists of two parts, prefix and suffix. The prefix is the set number $S_i$ in unary ($i$ 1's followed by a 0) and the suffix is the position of the codeword in the set (encoded in different ways, depending on the code). If a code satisfies this classification, then it can be described by (1) the size of set $i$ as a function of $i$ and (2) the encoding method of the suffix. The following are examples of UP/S codes:

1. The Rice code, where the parameter $m$ is a power of 2 ($m = 2^k$). Set $S_i$ consists of all the integers $n$ where $\lfloor n/2^k \rfloor$ equals $i$ (see the column labeled "No. of ones" in Table 2.50).

2. The start-step-stop codes. Start with $i = 0$ and increment $i$ by 1 to generate sets $S_i$. For each $i$, compute $a = \text{start} + i \times \text{step}$ and construct all the $a$-bit integers. There are $2^a$ such numbers and they become the suffixes of set $S_i$.

3. The exponential Golomb codes depend on a choice of a nonnegative integer parameter $s$. The discussion on page 165 mentions that the length of a codeword increases by 2 each time $1 + n/2^s$ is a power of 2. This naturally partitions the code into sets $S_i$, each including codewords for the integers from $2^i$ to $2^{i+1} - 1$.

> There is a surprising variety of codes
> (and some varieties of surprising codes).
>
> —Peter Fenwick

# Bibliography

Acharya, Tinku and Joseph F. JáJá (1995) "Enhancing LZW Coding Using a Variable-Length Binary Encoding," Tech Report TR 1995-70, Institute for Systems Research, University of Maryland.

Acharya, Tinku and Joseph F. JáJá (1996) "An On-Line Variable Length Binary Encoding of Text," *Information and Computer Science*, **94**:1–22.

Ahlswede, Rudolf, Te Sun Han, and Kingo Kobayashi (1997) "Universal Coding of Integers and Unbounded Search Trees," *IEEE Transactions on Information Theory*, **43**(2):669–682.

amu (2006) is http://ifa.amu.edu.pl/~kprzemek.

Anh, Vo Ngoc and Alistair Moffat (2005) "Inverted Index Compression Using Word-Aligned Binary Codes," *Information Retrieval*, **8**:151–166.

Apostolico, Alberto and A. S. Fraenkel (1987) "Robust Transmission of Unbounded Strings Using Fibonacci Representations," *IEEE Transactions on Information Theory*, **33**(2):238–245, March.

Bauer, Rainer and Joachim Hagenauer (2001) "On Variable-Length Codes for Iterative Source/Channel-Decoding," *Proceedings of the IEEE Data Compression Conference, 2001*, Snowbird, UT, pp. 273–282.

Bell, Timothy C., John G. Cleary, and Ian H. Witten (1990) *Text Compression*, Englewood Cliffs, NJ, Prentice Hall.

Bentley, J. L. and A. C. Yao (1976) "An Almost Optimal Algorithm for Unbounded Searching," *Information Processing Letters*, **5**(3):82–87.

Bass, Thomas A. (1992) *Eudaemonic Pie*, New York, Penguin Books.

Berger, T. and R. Yeung (1990) "Optimum '1'-Ended Binary Prefix Codes," *IEEE Transactions on Information Theory*, **36**(6):1435–1441, November.

Berstel, Jean and Dominique Perrin (1985) *Theory of Codes*, Orlando, FL, Academic Press.

Boldi, Paolo and Sebastiano Vigna (2004a), "The WebGraph Framework I: Compression Techniques," in *Proceedings of the 13th International World Wide Web Conference (WWW 2004)*, pages 595–601, New York, ACM Press.

Boldi, Paolo and Sebastiano Vigna (2004b), "The WebGraph Framework II: Codes for the World-Wide Web," in *Data Compression Conference, DCC 2004*.

Boldi, Paolo and Sebastiano Vigna (2005), "Codes for the World-Wide Web," *Internet Mathematics*, **2**(4):405–427.

Bookstein, Abraham and S. T. Klein (1993) "Is Huffman Coding Dead?" *Proceedings of the 16th Annual International ACM SIGIR Conference on Research and Development in Information Retrieval*, pp. 80–87. Also published in *Computing*, **50**(4):279–296, 1993, and in *Proceedings of the Data Compression Conference, 1993*, Snowbird, UT. p. 464.

Burrows, Michael and D. J. Wheeler (1994) *A Block-Sorting Lossless Data Compression Algorithm*, Digital Systems Research Center Report 124, Palo Alto, CA, May 10.

Buttigieg, Victor (1995a) *Variable-Length Error-Correcting Codes*, PhD thesis, University of Manchester.

Buttigieg, Victor and P. G. Farrell (1995b) "A Maximum A-Posteriori (MAP) Decoding Algorithm For Variable-Length Error-Correcting Codes," *Codes and Cyphers: Cryptography and Coding IV*, Essex, England, Institute of Mathematics and Its Applications, pp. 103–119.

Capocelli, Renato (1989) "Comments and Additions to 'Robust Transmission of Unbounded Strings Using Fibonacci Representations'," *IEEE Transactions on Information Theory*, **35**(1):191–193, January.

Capocelli, R. and A. De Santis (1992) "On The Construction of Statistically Synchronizable Codes," *IEEE Transactions on Information Theory*, **38**(2):407–414, March.

Capocelli, R. and A. De Santis (1994) "Binary Prefix Codes Ending in a '1'," *IEEE Transactions on Information Theory*, **40**(4):1296–1302, July.

Chaitin, Gregory J. (1966) "On the Lengths of Programs for Computing Finite Binary Sequences," *Journal of the ACM*, **13**(4):547–569, October.

Choueka Y., Shmuel T. Klein, and Y. Perl (1985) "Efficient Variants of Huffman Codes in High Level Languages," *Proceedings of the 8th ACM-SIGIR Conference*, Montreal, pp. 122–130.

codethatword (2007) is `http://www.codethatword.com/`.

dartmouth (2006) is
`http://www.math.dartmouth.edu/~euler/correspondence/letters/000765.pdf`.

Davisson, I. D. (1966) "Comments on 'Sequence Time Coding for Data Compression'," *Proceedings of the IEEE*, **54**:2010, December.

Elias, P. (1975) "Universal Codeword Sets and Representations of the Integers," *IEEE Transactions on Information Theory*, **21**(2):194–203, March.

Even, S. and M. Rodeh (1978) "Economical Encoding of Commas Between Strings," Communications of the ACM, 21(0)1418 317, April.

Fano, R. M. (1949) "The Transmission of Information," Research Laboratory for Electronics, MIT, Tech Rep. No. 65.

Fenwick, Peter (1996) "Punctured Elias Codes for Variable-Length Coding of the Integers," Technical Report 137, Department of Computer Science, University of Auckland, December. This is also available online at
`www.firstpr.com.au/audiocomp/lossless/TechRep137.pdf`.

Fenwick, P. (2002), "Variable-Length Integer Codes Based on the Goldbach Conjecture, and Other Additive Codes," *IEEE Transactions on Information Theory*, **48**(8):2412–2417, August.

Ferguson, T. J. and J. H. Rabinowitz (1984) "Self-Synchronizing Huffman codes," *IEEE Transactions on Information Theory*, **30**(4):687–693, July.

Fiala, E. R. and D. H. Greene (1989), "Data Compression with Finite Windows," *Communications of the ACM*, **32**(4):490–505.

firstpr (2006) is `http://www.firstpr.com.au/audiocomp/lossless/#rice`.

Fraenkel, A. S. and Shmuel T. Klein (1985) "Robust Universal Complete Codes as Alternatives to Huffman Codes," Tech. Report CS85-16, Dept. of Applied Mathematics, Weizmann Institute of Science, October.

Fraenkel, Aviezri S. and Shmuel T. Klein (1990) "Bidirectional Huffman Coding," *The Computer Journal*, **33**:296–307.

Fraenkel, Aviezri S. and Shmuel T. Klein (1996) "Robust Universal Complete Codes for Transmission and Compression," *Discrete Applied Mathematics*, **64**(1):31–55, January.

Freeman, G. H. (1991) "Asymptotic Convergence of Dual-Tree Entropy Codes," *Proceedings of the Data Compression Conference (DCC '91)*, pp. 208–217.

Freeman, G. H. (1993) "Divergence and the Construction of Variable-to-Variable-Length Lossless Codes by Source-Word Extensions," *Proceedings of the Data Compression Conference (DCC '93)*, pp. 79–88.

Gallager, Robert G. (1978) "Variations on a Theme by Huffman," *IEEE Transactions on Information Theory*, **24**(6):668–674, November.

Gallager, Robert G., and David C. van Voorhis (1975) "Optimal Source Codes for Geometrically Distributed Integer Alphabets," *IEEE Transactions on Information Theory*, **21**(3):228–230, March.

Gemstar (2006) is `http://www.gemstartvguide.com`.

Gilbert, E. N. and E. F. Moore (1959) "Variable Length Binary Encodings," *Bell System Technical Journal*, **38**:933–967.

Girod, Bernd (1999) "Bidirectionally Decodable Streams of Prefix Code-Words," *IEEE Communications Letters*, **3**(8):245–247, August.

Glusker, Mark, David M. Hogan, and Pamela Vass (2005) "The Ternary Calculating Machine of Thomas Fowler," *IEEE Annals of the History of Computing*, **27**(3):4–22, July.

Golomb, Solomon W. (1966) "Run-Length Encodings," *IEEE Transactions on Information Theory*, **12**(3):399–401.

Grimm, R. E. (1973) "The Autobiography of Leonardo Pisano," *Fibonacci Quarterly*, **11**(1):99–104, February.

Hirschberg, D. and D. Lelewer (1990) "Efficient Decoding of Prefix Codes," *Communications of the ACM*, **33**(4):449–459.

Howard, Paul G. (1998) "Interleaving Entropy Codes," *Proceedings Compression and Complexity of Sequences 1997*, Salerno, Italy, pp. 45–55, June.

Howard, Paul G. and J. S. Vitter (1994) "Fast Progressive Lossless Image Compression," Proceedings of the Image and Video Compression Conference, *IS&T/SPIE 1994 Symposium on Electronic Imaging: Science & Technology*, 2186, San Jose, CA, pp. 98–109, February.

Huffman, David (1952) "A Method for the Construction of Minimum Redundancy Codes," *Proceedings of the IRE*, **40**(9):1098–1101.

Karp, R. S. (1961) "Minimum-Redundancy Coding for the Discrete Noiseless Channel," *Transactions of the IRE*, **7**:27–38.

Kiely, A. (2004) "Selecting the Golomb Parameter in Rice Coding," IPN (Interplanetary Network) Progress Report, **42–159**:1–8, November 15.

Kraft, L. G. (1949) *A Device for Quantizing, Grouping, and Coding Amplitude Modulated Pulses*, Master's Thesis, Department of Electrical Engineering, MIT, Cambridge, MA.

Laković, Ksenija and John Villasenor (2002) "On Design of Error-Correcting Reversible Variable Length Codes," *IEEE Communications Letters*, **6**(8):337–339, August.

Laković, Ksenija and John Villasenor (2003) "An Algorithm for Construction of Efficient Fix-Free Codes," *IEEE Communications Letters*, **7**(8):391–393, August.

Lam, Wai-Man and Sanjeev R. Kulkarni (1996) "Extended Synchronizing Codewords for Binary Prefix Codes," *IEEE Transactions on Information Theory*, **42**(3):984–987, May.

Lawrence, John C. (1977) "A New Universal Coding Scheme for the Binary Memoryless Source," *IEEE Transactions on Information Theory*, **23**(4):466–472, July.

LBE (2007) is http://in.geocities.com/iamthebiggestone/how_lbe_works.htm.

Levenstein (2006) is http://en.wikipedia.org/wiki/Levenstein_coding.

Lynch, T. J. (1966) "Sequence Time Coding for Data Compression," *Proceedings of the IEEE*, **54**:1490–1491, October.

Mahoney, Michael S. (1990) "Goldbach's Biography" in Charles Coulston Gillispie *Dictionary of Scientific Biography*, New York, NY, Scribner's, 14 vols. 1970–1990.

McMillan, Brockway (1956) "Two Inequalities Implied by Unique Decipherability," *IEEE Transactions on Information Theory*, **2**(4).115–116, December.

morse-tape (2006) is `http://memory.loc.gov/mss/mmorse/071/071009/0001d.jpg`.

pass (2006) is `http://pass.maths.org/issue2/xfile/`.

Pigeon, Steven (2001a) "Start/Stop Codes," *Proceedings of the Data Compression Conference (DCC '01)*, p. 511. Also available at
`http://www.stevenpigeon.org/Publications/publications/ssc_full.pdf`.

Pigeon, Steven (2001b) *Contributions to Data Compression*, PhD Thesis, University of Montreal (in French). Available from
`http://www.stevenpigeon.org/Publications/publications/phd.pdf`. The part on taboo codes is "Taboo Codes, New Classes of Universal Codes," and is also available at `www.iro.umontreal.ca/~brassard/SEMINAIRES/taboo.ps`. A new version has been submitted to *SIAM Journal of Computing*.

Randall, Keith, Raymie Stata, Rajiv Wickremesinghe, and Janet L. Wiener (2001) "The LINK Database: Fast Access to Graphs of the Web." *Research Report 175*, Compaq Systems Research Center, Palo Alto, CA.

Rice, Robert F. (1979) "Some Practical Universal Noiseless Coding Techniques," Jet Propulsion Laboratory, JPL Publication 79-22, Pasadena, CA, March.

Rice, Robert F. (1991) "Some Practical Universal Noiseless Coding Techniques—Part III. Module PSI14.K," Jet Propulsion Laboratory, JPL Publication 91-3, Pasadena, CA, November.

Robinson, Tony (1994) "Simple Lossless and Near-Lossless Waveform Compression," Technical Report CUED/F-INFENG/TR.156, Cambridge University, December. Available at `http://citeseer.nj.nec.com/robinson94shorten.html`.

Rudner, B. (1971) "Construction of Minimum-Redundancy Codes with an Optimum Synchronization Property," *IEEE Transactions on Information Theory*, **17**(4):478–487, July.

> "So few?" asked Frost. "Many of them contained bibliographies of books I have not yet scanned." "Then those books no longer exist," said Mordel. "It is only by accident that my master succeeded in preserving as many as there are."
>
> —Roger Zelazny, *For a Breath I Tarry* (1966)

Salomon, D. (2003) *Data Privacy and Security*, New York, Springer Verlag.

Salomon, D. (2006) *Data Compression: The Complete Reference*, 4th edition, London, Springer Verlag.

Schalkwijk, J. Pieter M. "An Algorithm for Source Coding," *IEEE Transactions on Information Theory*, **18**(3):395–399, May.

seul.org (2006) is
http://f-cpu.seul.org/whygee/phasing-in_codes/PhasingInCodes.nb.

Shannon, Claude E. (1948), "A Mathematical Theory of Communication," *Bell System Technical Journal*, **27**:379–423 and 623–656, July and October,

Sieminski, A. (1988) "Fast Decoding of the Huffman Codes," *Information Processing Letters*, **26**(5):237–241.

Sloane, Neil J. A. (2006) "The On-Line Encyclopedia of Integer Sequences," at www.research.att.com/~njas/sequences/.

Stout, Quentin F. (1980) "Improved Prefix Encodings of the Natural Numbers," *IEEE Transactions on Information Theory*, **26**(5):607–609, September.

T-REC-h (2006) is http://www.itu.int/rec/T-REC-h.

Takishima, Y., M. Wada, and H. Murakami, (1995) "Reversible Variable-Length Codes," *IEEE Transactions on Communications*, **43**(2,3,4):158–162, Feb./Mar./Apr.

Teuhola, J. (1978) "A Compression Method for Clustered Bit-Vectors," *Information Processing Letters*, **7**:308–311, October.

Tjalkens, T. and Frans M. Willems (1992) "A Universal Variable-to-Fixed Length Source Code Based on Lawrence's Algorithm," *IEEE Transactions on Information Theory*, **38**(2):247–253, March.

Tsai, C. W. and J. L. Wu (2001a) "On Constructing the Huffman-Code Based Reversible Variable Length Codes," *IEEE Transactions on Communications*, **49**(9):1506–1509, September.

Tsai, Chien-Wu and Ja-Ling Wu (2001b) "Modified Symmetrical Reversible Variable-Length Code and its Theoretical Bounds," *IEEE Transactions on Information Theory*, **47**(6):2543–2548, September.

Tunstall, B. P. (1967) "Synthesis of Noiseless Compression Codes," PhD dissertation, Georgia Institute of Technology, Atlanta, GA, September.

uklinux (2007) is http://www.bckelk.uklinux.net/menu.html.

> Outside of a dog, a book is man's best friend. Inside of a dog it's too dark to read.
>
> —Groucho Marx

Ulam (2006) is Weisstein, Eric W. "Ulam Sequence." From MathWorld–A Wolfram Web Resource. http://mathworld.wolfram.com/UlamSequence.html.

utm (2006) is http://primes.utm.edu/notes/conjectures/.

Wang, Muzhong (1988) "Almost Asymptotically Optimal Flag Encoding of the Integers," *IEEE Transactions on Information Theory*, **34**(2):324–326, March.

Wen, Jiangtao and John D. Villasenor (1998) "Reversible Variable Length Codes for Efficient and Robust Image and Video Coding," *Data Compression Conference*, pp. 471–480, Snowbird, UT, March–April.

Yamamoto, Hirosuke (2000) "A New Recursive Universal Code of the Positive Integers," *IEEE Transactions on Information Theory*, **46**(2):717–723, March.

Yamamoto, Hirosuke and Hiroshi Ochi (1991) "A New Asymptotically Optimal Code for the Positive Integers," *IEEE Transactions on Information Theory*, **37**(5):1420–1429, September.

Zeckendorf, E. (1972) "Représentation des Nombres Naturels par Une Somme de Nombres de Fibonacci ou de Nombres de Lucas," *Bull. Soc. Roy. Sci. Liège*, **41**:179–182.

Zeilberger, D. (1993) "Theorems for a Price: Tomorrow's Semi-Rigorous Mathematical Culture," *Notices of the American Mathematical Society*, **40**(8):978–981, October. Reprinted in *Mathematical Intelligencer*, **16**(4):11–14 (Fall 1994).

Zipf's Law (2007) is http://en.wikipedia.org/wiki/Zipf's_law

Ziv, Jacob and A. Lempel (1977) "A Universal Algorithm for Sequential Data Compression," *IEEE Transactions on Information Theory*, **23**(3):337–343.

Ziv, Jacob and A. Lempel (1978) "Compression of Individual Sequences via Variable-Rate Coding," *IEEE Transactions on Information Theory*, **24**(5):530–536.

Bibliography, in its most general sense is the study and description of books. It can be divided into enumerative or systematic bibliography, which results in an overview of publications in a particular category, and analytical or critical bibliography, which studies the production of books.

http://en.wikipedia.org/wiki/Bibliography

# Index

The index caters to those who have already read the book and want to locate a familiar item, as well as to those new to the book who are looking for a particular topic. I have included any terms that may occur to a reader interested in any of the topics discussed in the book (even topics that are just mentioned in passing). As a result, even a quick glance over the index gives the reader an idea of the terms and topics included in the book. Any errors and omissions brought to my attention are welcome. They will be added to the errata list and will be included in any future editions.

An index is a tool that simplifies the
measurement of movements in a numerical series.
http://www.bls.gov/cpi/cpifaq.htm

# Colophon

The idea for this little book came to me gradually during the summer of 2006, while I was working on the 4th edition of the much bigger book *Data Compression: The Complete Reference*. Searching the Internet for new compression methods and techniques, I came across several references to a document by Steven Pigeon on variable-length codes, but the document itself seemed to have evaporated. I became intrigued by the absence of this document, and I gradually realized that my book, big as it was, did not say much about these important codes. Once the big book was completed and was sent to the publisher, I decided to look deeper into variable-length codes. My original intention was to write perhaps 30–40 pages and leave the material, in PDF format, in the web site of the big book, as auxiliary material. However, once I started to actually collect information on the many variable-length codes and their applications, the document grew quickly and eventually caught the eye of Wayne Wheeler, the computer science editor of Springer Verlag, who decided to publish it as a book.

The material for this book was written as separate sections, from September 2006 to January 2007. It took a few more weeks to reorder the sections, read and reread the text, make improvements and corrections, and add the Preface, Introduction, and Index. The final manuscript was sent to the publisher in late February 2007.

The book is small and it consists of about 92,000 words. There are 170 cross-references (for figures, tables, equations, sections, and chapters) and the raw index file contains about 560 index items. Like all my previous books, this one was also typeset in plain TEX, with special formatting macros, using the Textures implementation on the Macintosh.

> Truly the gods have not from the beginning revealed all things to mortals, but by long seeking, mortals make progress in discovery.
>
> —Xenophanes of Colophon